Elevator Troubleshooting & Repair

Elevator Troubleshooting & Repair

A Technician's Certification Study Guide

David Herres

INDUSTRIAL PRESS, INC.

Industrial Press, Inc.

1 Chestnut Street
South Norwalk, Connecticut 06854
Phone: 203-956-5593
Toll-Free in USA: 888-528-7852
Email: info@industrialpress.com

Author: David Herres
Title: Elevator Troubleshooting & Repair: A Technician's Certification Study Guide
Library of Congress Control Number: 2020930841

© by Industrial Press.
All rights reserved. Published in 2020.

ISBN (print):	978-0-8311-3643-7
ISBN (ePDF):	978-0-8311-9527-4
ISBN (ePUB):	978-0-8311-9528-1
ISBN (eMOBI):	978-0-8311-9529-8

Editorial Director/Publisher: Judy Bass
Copy Editor: Janice Gold
Compositor: Patricia Wallenburg, TypeWriting
Proofreader: Alison Shurtz
Indexer: WordCo. Indexing Services, Inc.

books.industrialpress.com
ebooks.industrialpress.com

CONTENTS

History   1

Types of Elevators   21

CHAPTER 3

AC and DC Electric Motor Maintenance, VFD Troubleshooting, and Diagnostic Procedures   41

CHAPTER 4

Advanced Motor Repair   63

CHAPTER 5

Troubleshooting Elevator Systems   85

CHAPTER 10

Systems Connected to Elevator Installations and How They Work in Concert 199

APPENDIX A

Study Questions Answers 219

APPENDIX B

Electrical Laws and Equations 221

INTRODUCTION

An appropriate subtitle for this introduction would be: DON'T GET IN OVER YOUR HEAD! Elevators move large numbers of people up and down in high-rise buildings on a daily basis and there is the potential for an accident to occur. A highly-developed system of safeguards is in place, and for this reason an elevator ride is statistically far less hazardous than crossing a street.

Fatalities are extremely rare due to the redundant safety mechanisms, but they have occurred, some because of faulty maintenance procedures. A documented instance involved two wires which were temporarily disconnected from a circuit board and then reversed, thus disabling the door interlock, which prevents an elevator car from moving when the door is not closed and latched. A child was crushed between the car floor and door opening. There are tremendous moral and legal issues, and it is the responsibility of each worker to understand the consequences of any errors.

In the popular imagination, if an elevator cable breaks, the car immediately falls to the bottom of the shaft, killing the occupants. In actuality, the car is connected by multiple cables, which are regularly inspected for wear, any one of which would hold a fully-loaded car. Additionally, elevator cars have mechanisms known as safeties, which clamp onto the guide rails in an overspeeding or slack-cable situation.

In performing elevator maintenance and repair, there are enormous moral and legal issues. Workers need to fully understand proper maintenance procedures so that all safeguards remain in effect. As a start, it is essential to become aware of applicable regulations and to comply with them.

Of course, a major aspect in elevator technology is electrical. The National Electrical Code (NEC), revised and issued every three years by the National Fire Protection Association (NFPA), governs most electrical installations with a few exceptions such as installations under the direct control of utilities, and underground in mines. (It does, however, govern non-mine installations such as lighting and signal wiring in underground traffic tunnels.) NEC has no legal standing on its own, but is offered up for adoption and enforcement by states, municipalities, and jurisdictions inside and outside the United States.

NEC Article 620, part of Chapter 6, covers elevators, dumbwaiters, escalators, moving walks, platform lifts, and stairway lifts. Each of these performs a different set of functions with different requirements. The section on elevators contains specific electrical requirements, which must be observed in new installations. It is not the intent of NEC that with each Code revision all existing installations are to be immediately upgraded to comply. Nevertheless, older installations should be critically evaluated to see where upgrades are feasible and/or warranted. Besides Article 620, new elevator installations must comply with the entire electrical code except where specifically exempted. Two very important articles are 250, Grounding, and 430, Motors, Motor Circuits and Controllers.

I will return to the very important subject of codes in this Introduction and in fact throughout the book, but first a few paragraphs about licensing.

If you are serious about engaging in elevator work, you should acquire the appropriate license(s)—electrician's license and elevator mechanic's license. This is not going to happen all at once. Most states require verified work experience and/or completion of classroom or online training, including passage of an exam, plus the state exam. There is no single nationwide electrician's or elevator technician's license. For the most part in the United States these permits are issued and regulated by the individual states, or in a few instances jurisdiction is ceded over to separate counties or municipalities. Most states maintain electrical and elevator agencies, which inspect installations and issue technicians' and other permits. Requirements vary widely from state to state. The best approach as a start is to check your state's website for requirements and procedures for obtaining the appropriate licenses.

My home state, New Hampshire, requires electricians' licensing for those who are performing electrical installations for heat, light, and power purposes regardless of the voltage. It is not the voltage of the circuit that determines the requirements of licensure, it is the type of circuit. Accordingly, an individual can work on a central fire alarm system without having an electrician's license, because the electrical system is for signaling, not heat, light, or power. How does this apply to elevator work? Most inspectors, technically known as the authorities having jurisdiction, will recognize that elevators, as we shall see in the chapters that follow, are composed of both signaling and power electrical circuits. Therefore, the electrician's license would not be required to work on the signaling and control circuitry, in the car, on individual floors where the call buttons are located, in the machine room, or in the motion controller. However, the electrician's license would be required to work on the power path from the entrance panel, through the main disconnect, to the variable frequency drive (VFD) and motor. The electrician's license would also be required to work on lighting and branch-circuit receptacles in the pit and machine room. (In most cases a licensed electrician would in fact work on the signal wiring, but for this the license would not absolutely be required.) The bottom line is that if you are serious about doing elevator work, you should obtain the electrician's license.

The New Hampshire Board licenses three categories: master, journeyman, and high/medium voltage electricians. In addition, it registers apprentice electricians and high/medium voltage trainees. All electrical work for which an electrician's license is required must be overseen by a licensed master electrician. This individual need not be at the jobsite at all times, but is responsible for seeing that the work complies with the National Electrical Code. Journeyman electricians may perform the on-site electrical work, but they may not work on their own as independent contractors. Apprentice electricians may also perform electrical work, but they must be supervised by journeyman electricians, who must be on-site whenever apprentices are working. A one-on-one ratio must be maintained. If there are two apprentice electricians performing electrical work, there must be two journeyman or master electricians at the site.

To receive a master electrician's license, the individual must have 8000 hours practical experience as an apprentice to a licensed electrician and at least one of the following:

- 575 hours of electrical schooling in blocks of 144 hours per year or have an associate or higher degree in an electrical curriculum
- Have ten years experience as a journeyman or master electrician as required in another jurisdiction; or
- Have taken the journeyman or master exam in New Hampshire previously
- Acquired credit or school time, not to exceed 2000 hours, towards the practical experience requirement or completion of prescribed courses in electrical installations at an approved school
- Pass the journeyman examination and obtain 2000 hours of experience as a journeyman in performing electrical installations prior to being examined

The master electrician exam consists of 50 NEC questions, 50 questions on practical installations, and 25 questions on applicable state laws and the Board's administrative rules. The license fee is $270 for three years. To receive a journeyman electrician's license, the requirements are substantially the same, but the exam is less difficult. The fee is $150 for three years.

The apprentice is not required to take an examination or have work experience prior to application, but must possess a high school diploma or equivalent. The fee for the apprentice card is $30, and it is valid for one year. During that period the apprentice is expected to complete 150 hours of vocational training and to study for the journeyman's license. New Hampshire also issues an elevator mechanic's license. Considerable practical experience working with a licensed individual is required to obtain this license, so it would be a long-range goal.

The American Society of Mechanical Engineers (ASME), through its Board on Safety Codes and Standards (BSCS), develops and maintains a comprehensive portfolio of codes and standards that governs elevators and escalators. A complete listing can be seen at ASME.org.

ASME A17, Safety Code for Existing Elevators and Escalators, is an essential reference for elevator maintenance workers and repair technicians. It can be ordered online at ASME.org. After a few paragraphs on Purpose and Exceptions, ASME A17 lists pertinent definitions. Here, by way of overview, are some highlights, with references to escalators, moving walks, and material lifts eliminated because they are outside the scope of this textbook.

- **Car annunciator:** An electrical device in the car that indicates visually the landings at which an elevator landing signal registering device has been actuated.
- **Auxiliary power lowering device:** An alternatively powered auxiliary control system that will, upon failure of the main power supply, allow a hydraulic elevator to descend to a lower landing.
- **Emergency brake:** A mechanical device independent of the braking system used to retard or stop an elevator should the car overspeed or move in an unintended manner. Such devices include, but are not limited to, those that apply braking force on car rails, counterweight rails, suspension or compensation ropes, drive sheaves, and brake drums.
- **Buffer:** A device designed to stop a descending car or counterweight beyond its normal limit of travel by storing or by absorbing and dissipating the kinetic energy of the car or counterweight.
- **Bumper:** A device, other than an oil or spring buffer, designed to stop a descending car or counterweight beyond its normal limit of travel by absorbing the impact.
- **Car-direction indicator:** A visual signaling device that displays the current direction of travel.
- **Car door interlock:** A device having two related and interdependent functions, which are to prevent the operation of the driving machine by the normal operating device unless the car door is locked in the closed position; and to prevent the opening of the car door from inside the car unless the car is within the landing zone and is either stopped or being stopped.
- **Car door or gate electric contact:** An electrical device, the function of which is to prevent operation of the driving machine by the normal operating device unless the car door or gate is in the closed position.
- **Overslung car frame:** A car frame to which the hoisting rope fastenings or hoisting rope sheaves are attached to the crosshead or top member of the car frame.
- **Underslung car frame:** A car frame to which the hoisting rope fastenings or hoisting rope sheaves are attached at or below the car platform.
- **Car lantern:** An audible and visual signaling device located in a car to indicate the car is answering the call and the car's intended direction of travel.
- **Car platform:** The structure that forms the floor of the car and directly supports the load.
- **Compensating rope sheave switch:** A device that automatically causes the electric power to be removed from the elevator driving-machine motor and brake when the compensating sheave approaches its upper or lower limit of travel.

- **Motion control:** That portion of a control system that governs the acceleration, speed, retardation, and stopping of the moving member.
- **Operation control:** That portion of a control system that initiates the starting, stopping, and direction of motion in response to a signal from an operating device.
- **Automatic operation:** Operation control wherein the starting of the elevator car is effected in response to the momentary actuation of operating devices at the landing, and/or of operating devices in the car identified with the landings, and/or in response to an automatic starting mechanism, and wherein the car is stopped automatically at the landings.
- **Motion controller:** An operative unit comprising a device or group of devices for actuating the moving member.
- **Motor controller:** The operative units of a motion control system comprising the starter devices and power conversion equipment required to drive an electric motor.
- **Elevator control room:** An enclosed control space outside the hoistway, intended for full bodily entry, which contains the motor controller. The room could also contain electrical and/or mechanical equipment used directly in connection with the elevator, but not the electric driving machine or the hydraulic machine.
- **Control system:** The overall system governing the starting, stopping, direction of motion, acceleration, speed, and retardation of the moving member.
- **Designated level:** The main floor or other floor level that best serves the needs of emergency personnel for firefighting or rescue purposes identified by the building code or fire authority.
- **Displacement switch:** A device actuated by the displacement of the counterweight, at any point in the hoistway, to provide a signal that the counterweight has moved from its normal lane of travel or has left its guide rails.
- **Door or gate electric contact:** An electrical device, the function of which is to prevent operation of the driving machine by the normal operating device unless the door or gate is in the closed position.
- **Electrical/electronic/programmable electronic (E/E/PE):** Based on electrical (E) and/or electronic (E) and/or programmable electronic (PE) technology.
- **Hydraulic elevator:** A power elevator in which the energy is applied, by means of a liquid under pressure, in a hydraulic jack.
- **Roped hydraulic elevator:** A hydraulic elevator in which the energy is applied by a roped-hydraulic driving machine.
- **Hoistway door or gate locking device:** A device that secures a hoistway door or gate in the closed position and prevents it from being opened from the landing side except under certain specified conditions.
- **Hydraulic jack:** A unit consisting of a cylinder equipped with a plunger (ram) or piston, which applies the energy provided by a liquid under pressure.

- **Driving machine:** The power unit that applies the energy necessary to drive an elevator.
- **Geared driving machine:** A direct driving machine in which the energy is transmitted from the motor to the driving sheave, drum, or shaft through gearing.
- **Traction machine:** A direct driving machine in which the motion of a car is obtained through friction between the suspension ropes and a traction sheave.
- **Gearless traction machine:** A traction machine, without intermediate gearing, that has the traction sheave and the brake drum mounted directly on the motor shaft.
- **Hydraulic driving machine:** A driving machine in which the energy is provided by a hydraulic machine and applied by a hydraulic jack.
- **Direct hydraulic driving machine:** A hydraulic driving machine in which the driving member of the hydraulic jack is directly attached to the car frame or platform.
- **Roped-hydraulic driving machine:** A hydraulic driving machine in which the driving member of the hydraulic jack is connected to the car by wire ropes or indirectly coupled to the car by means of wire ropes and sheaves. It includes multiplying sheaves, if any, and their guides.
- **Elevator machine room:** An enclosed machinery space outside the hoistway, intended for full bodily entry, which contains the electric driving machine or the hydraulic machine. The room could also contain electrical and/or mechanical equipment used directly in connection with the elevator.
- **Maintained pressure:** The hydraulic pressure between the pressure source and the control valves of a maintained pressure hydraulic elevator.
- **Compensation means:** The method by which unbalanced forces due to suspension means are reduced, utilizing one or more compensation members and their terminations.
- **Elevator nonstop switch:** A switch that, when operated, will prevent the elevator from making registered landing stops.
- **Inspection operation:** A special case of continuous-pressure operation used for troubleshooting, maintenance, repair, adjustments, rescue, and inspection.
- **Phase I recall operation:** The operation of an elevator where it is automatically or manually recalled to the recall level and removed from normal service because of activation of firefighters' emergency operation.
- **Phase II emergency in-car operation:** The operation of an elevator by firefighters where the elevator is under their control.
- **Elevator pit:** The portion of a hoistway extending from the sill level of the bottom terminal landing to the floor at the bottom of the hoistway.
- **Rated speed:** The speed at which an elevator is designed to operate.
- **Car or counterweight safety:** A mechanical device attached to the car, car frame, or to an auxiliary frame, or to the counterweight or counterweight frame in order to stop and hold the counterweight under one or more of the

following conditions: predetermined overspeed, free fall, or if the suspension ropes slacken.

- **Traveling cable:** A cable made up of electric conductors, which provides electrical connection between an elevator car or counterweight and a fixed outlet in the hoistway or machine room.
- **Unintended car movement:** Any movement of an elevator car that is not intended car movement, resulting from a component or system failure.

Besides the foregoing definitions, which give us a valuable overview, Part II (Hoistways and Related Construction for Electric Elevators), Part III (Machinery and Equipment for Electric Elevators) and Part IV (Hydraulic Elevators) provide specific mandates and construction details. It is necessary to comply with these ASME A17 provisions in order to ensure that existing installations remain safe for workers and users. In performing repairs, it is essential that safety mechanisms remain in place, intact and fully functional. Moreover, workers should constantly evaluate the overall installation in the context of NEC and ASME A17 compliance. Additionally, Occupational Safety and Health Administration (OSHA) regulations are relevant to workplace safety. Keep in mind that most elevator accidents involve not members of the public who are using the equipment, but workers who are installing, maintaining, and repairing it. The obvious hazards are falling, being crushed, and being electrocuted. Additionally, there are long-term health hazards. Proper workplace procedures protect technicians and workers.

ASME A17 Part II applies to hoistways and related construction for electric elevators. It begins with the very basic statement that hoistways are to be enclosed throughout their height. The enclosure protects elevator machinery to a limited but very significant extent from fire that may occur in an adjacent area. This buys time for elevator occupants to descend to the ground floor (Phase I) and for firefighters to take manual control of the elevator (Phase II) so that they can endeavor to halt the fire and perform rescue operations.

Notwithstanding the fact that the hoistway protects the car and occupants from an external fire, the fact must be recognized that in some instances this vertical shaft can act as a highly efficient flue, increasing a fire's temperature and facilitating its spread to upper floors. We shall see in Chapter 1, History, some attempts to confront this problem, for example by installing safety hatches, which consisted of sliding or hinged panels at each floor to create a series of closed compartments to control the draft. These panels would open to allow the car to pass. The hoistway moreover serves as a supporting structure for cable sheaves, counterweight guards, doorways at landings, control wiring, guide rails, and safeties.

Protection is to be provided around elevators adjacent to areas permitting passage of people and adjacent to areas permitting storage. This protection is permitted to be fixed guards, or sufficient distance from the moving portion of the elevator, or a combination of both, so that no one can accidentally come in contact with the elevator. Hoistway enclosures must have substantially flush surfaces on the hoistway sides used for loading and unloading. Landing sills, hoistway doors, door tracks, and hangers are permitted to project inside the hoistway enclosure.

- Where a car leveling device is provided and the hoistway edge of the sill is either flush with or projects into the hoistway, the guard is to have a straight vertical face extending below the sill not less than the leveling zone plus three inches.
- Where the sill projects inward from the hoistway enclosure, the bottom of the guard is to be also beveled at an angle of not less than 60 degrees nor more than 75 degrees from the horizontal or the guard is to be extended from the hoistway edge of the landing sill to the top of the door hanger pocket of the next entrance below.
- Where no car leveling device is provided and the sill projects inward from the general line of the hoistway, the guard is to be either beveled at an angle of not less than 60 degrees nor more than 75 degrees from the horizontal, or it is permitted to have a straight vertical face extending from the hoistway edge of the landing sill to the top of the door hanger pocket of the next entrance below.
- Metal guards are to be installed in the pit and/or machine room located underneath the hoistway on all open sides of the counterweight runway except that where a compensating chain(s) or rope(s) is attached to the counterweight; the guard is permitted to be omitted in the pit on the side facing the elevator car to which these chains or ropes are attached.
- Where pit-mounted buffers are used, the guard is permitted to be omitted where the bottom of the counterweight resting on its compressed buffer is seven feet or more above the pit floor or above the machine or control room floor if located under the hoistway.
- A permanent means of access to elevator machine rooms and machinery spaces is to be provided for authorized persons. Access doors to machine rooms and machinery spaces are to be kept closed and locked. The only means of access to a machine room is not to be through the hoistway. Permanent lighting is to be provided in all machine rooms and machinery spaces.
- Means of access for authorized personnel is to be provided to all pits. A stop switch is to be provided in the pit for every elevator.
- For hoistway doors, interlocks are required for passenger elevators.
- Elevators that are operated from within the car are to have elevator parking devices installed at every landing that is equipped with an unlocking device.

ASME A17 Part III covers machinery and equipment for electric elevators.

- Car and counterweight buffers or bumpers are to be provided. Solid bumpers may be permitted in lieu of buffers.
- On rod-type counterweights, the rod nuts are to be cotter-pinned and the tie rods are to be protected so that head weight cannot crush the tie rods on buffer engagement. The weights are to be protected so they cannot be dislodged.

- Every elevator car is to have a platform consisting of a non-perforated floor attached to the platform frame supported by the car frame, and extending over the entire area within the car enclosure.
- Hinged platform sills, where provided, are to have electric contacts that will prevent operation of the elevator by a normal operating device unless the hinged sill is within two inches of its fully retracted position.
- The elevator is permitted to be operated by the leveling device in the leveling zone with the sill in any position.
- Floating (movable) platforms that permit operation of the elevator when the car door or gate is not in the closed position are prohibited.
- Cars are to be fully enclosed on all non-entrance sides and on top.
- Cars are to have a car door or gate provided at each entrance equipped with a car door or gate electric contact. It is to be positively opened by a lever or other device attached to and operated by the door or gate.
- For elevators installed in enclosed hoistways, cars are to be provided with a car top emergency exit with a cover hinged or otherwise attached to the car top so that the cover can be opened from the top of the car only, and opens outward. Interiors of cars are to be provided with electric light or lights. No less than two lamps are to be provided.
- The car of every elevator suspended by wire ropes is to be provided with a safety capable of stopping and sustaining the car with rated load. When the safety is operated by a governor, the safety is to be capable of stopping and sustaining the car with rated load from governor tripping speed. Counterweight safeties are to be provided and are to be capable of stopping and sustaining the counterweight.
- Safeties are to be applied mechanically. Electrical, hydraulic, or pneumatic devices are not to be used to apply the safeties nor to hold the safeties in the retracted position. Safeties that depend upon traction for application are prohibited. When car safeties are applied, no decrease in tension in the governor rope nor motion of the car in the down direction is permitted to release the safeties, but the safeties are permitted to be released by the motion of the car in the up direction.
- Rail lubricants or coatings that will reduce the holding power of the safety or prevent its functioning as required are not to be used.
- The car and counterweight guide rails are to extend at the top and bottom to prevent the guiding member from disengaging from the guide rails in the event that either the car or counterweight reaches its extreme limit of travel.
- Sheaves and drums are to be of cast iron or steel and are to have finished grooves for ropes.
- Winding drum machines are to be provided with a slack rope device having an enclosed switch of the manually reset type that will cause the electric power to be removed from the elevator driving-machine motor and brake if the hoisting ropes become slack or broken.

- In indirect-drive machines, each chain or belt in a set is to be continuously monitored by a broken belt or chain device of the manually reset type, which will function to automatically interrupt power to the machine and apply the brake in the event any belt or chain in the set breaks or becomes excessively slack. If one belt or chain of a set is worn, stretched, or damaged so as to require replacement, the entire set is to be replaced. Sprockets and toothed sheaves are also to be inspected on such occasions and to be replaced if noticeably worn.
- The elevator-driving machine is to be equipped with a friction brake applied by a spring or springs, or by gravity, and is to be released electrically. The brake is to be designed to have a capacity sufficient to hold the car at rest with its rated load.
- Enclosed upper and lower normal stopping devices are to be provided and arranged to slow down and stop the car automatically, at or near the top and bottom terminal landings. These devices are to function independently of the operation of the normal stopping means and of the final terminal stopping device.
- Manually operated rope (shipper rope) or rod-operating devices, or rope-operating devices actuated by wheels, levers, or cranks are not to be used.
- Handles of lever-type operating devices of car-switch operation elevators are to be so arranged that they will return to the stop position and latch there automatically when the hand of the operator is removed.
- Elevators with automatic or continuous-pressure operations are to have a continuous pressure button operating switch mounted on the top of the car for the purpose of operating the car solely from the top of the car. The device is to operate the car at a speed not exceeding 150 feet per minute.
- The means for transferring the control of the elevator to the top-of-car operating device is to be on the car top and located between the car crosshead and the side of the car nearest the hoistway entrance normally used for access to the car top.

Electrical Protective Devices, as covered in ASME A17, Part III, will be discussed in detail in Chapter 5, Troubleshooting Elevator Systems.

Hydraulic elevators, as covered in ASME A17, Part IV, will be discussed in detail in Chapter 2, Types of Elevators.

The requirements listed above are a small part of the entire elevator Code. They are presented by way of introduction to the general topics covered in this book.

ACKNOWLEDGMENTS

Many thanks to:

- Judy Bass, Publisher, Industrial Press Inc., who envisioned this work and suggested viable approaches
- Janice Gold, copy editor, and Patricia Wallenburg, compositor, who together wrote the book on competency and accuracy
- Judi Howcroft, supreme nature and industrial photographer, who showed me the way, departed, and lives on among the stars
- Deidre Schardine, an ongoing inspiration, who knows what is right and how to get there

Elevator Troubleshooting
& Repair

CHAPTER 1

HISTORY

It is likely that animal- and human-powered elevators predated written history. Unlike masonry and stone buildings, the cars were probably woven baskets or wooden platforms with or without guardrails, and the support structures built of wooden logs, so these remains would have decayed centuries ago. We can only surmise that they existed, powered by domesticated animals on the ground, who worked long hours at a turnstile. Alternatively, occupants of the car may have pulled a looped rope that turned a pulley with more ropes that lifted the car, as shown in Figure 1-1.

FIGURE 1-1 The rope was operated from within the car. The hoistway was primitive, but it did the job.

Early Elevators

Vitruvius (c. 80–15 bc), a Roman author, architect, and engineer, provided the first extant written reference. He reported that the Greek mathematician Archimedes (c. 287–212 bc) built a bank of elevators operated by hoisting ropes wrapped about a drum. It was turned by humans and this torque was applied to a capstan, causing platforms to lift gladiators and fierce animals through vertical shafts into the arena. In the seventeenth century, English and French monarchs built "flying chairs" to discreetly transport their mistresses to upper palace levels. These machines, powered by humans and animals, were eventually eclipsed by steam, water, and finally electric motors.

Where it gets interesting, from our point of view, is in the nineteenth century. During this 100-year period, the elevator evolved from steam-powered platforms used to move coal in English mines, to electrically-powered elevators that lifted passengers to ever greater heights in comfortable rooms with plush furniture.

In the late 1790s, William Strutt (1756–1830), shown in Figure 1-2, assumed control of his father's textile mills in England. Among many projects, including fireproofing and improving the heat system, he designed a combination passenger and freight elevator, known then as the crane. It was adjacent to the main stairway and was used to transport workers within the five-story building. Strutt's elevator was powered by a flat belt, running off of power shafting that ran throughout the building, presumably powered by an outside water wheel.

The principle components were a brake wheel, two fixed and two free pulleys, two endless belts, and a belt shifter. A crossed belt permitted the direction of car motion to be reversed, as needed in any elevator.

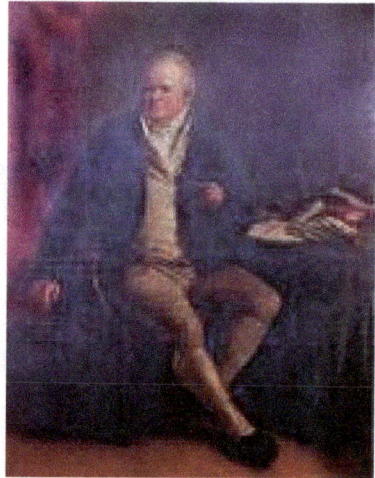

FIGURE 1-2 William Strutt (1756–1830) (*Wikipedia*)

A pinion gear was attached to one end of the main shaft, and its teeth meshed with those of a spur gear attached to the hoisting pulley shaft.

This was the first in a long series of working elevators that spanned the nineteenth century. Strutt's Teagle, as it was known, was complex in the sense that it had a lot of ropes, belts, and pulleys, but simple in that these things worked smoothly together to deliver the power to where it was needed so that the car could deliver workers throughout Strutt's five-story textile mill.

By the 1840s, two trends in vertical transportation merged. Increasingly, elevators were optimized to carry freight exclusively or to transport only workers, residents of tall buildings, and hotel guests from ground level to the growing number of floors in taller buildings that began to crowd the cities. Also, of necessity there was greater emphasis on safety.

Safety Measures

Previously, lower-powered lifting machines had their share of accidents, sometimes resulting in well-publicized fatalities. This was true not only in elevators, but throughout the world of increasingly mechanized, more powerful and faster machinery that characterized the new industrial age. Accidents took two forms. In one, the suspension rope and associated rigging that raised and lowered the car in a traction elevator failed, causing the car, which was slowed only a little by the air column below, to free fall to the bottom of the shaft. The inevitable result was severe injury, often fatal. The other type of accident involved the absence of reliable door interlocks, which would prevent a door from opening when the car was moving and/or prevent the car from moving when the doors were not closed and locked.

Without these interlocks, an occupant of the building could step through an open door assuming that the car was at the landing, and fall to the bottom of the shaft. Another equally great hazard was that an occupant of the car could be crushed between the car floor and the top of the door opening at any floor while the car was ascending. We shall see how mid-nineteenth century advances in elevator technology confronted these hazards and greatly reduced the number of injuries resulting from them.

Before midcentury, freight elevators were typically designed in-house to meet the needs of the many industrial facilities that were appearing, especially in England and eastern U.S. Then, beginning around 1845, industries and commercial operations such as hotels and office buildings began to look to certain emerging elevator manufacturers to meet these needs. Henry Waterman in New york City was a freight and passenger elevator manufacturer. One of his early machines, built in Manhattan for Croton Flour Mills, was operated from within the car so that an outside attendant was not required. Car motion was initiated by moving a simple iron lever, rather than tugging on the shipper rope. For passengers, the trip became smoother and more user-friendly. The control lever moved an attached chain that passed through openings in the car roof and floor, then engaged devices at the top of the shaft. The mechanism consisted of a friction clutch driven by a conventional power shaft, eliminating the need for pulleys and a belt shifter as in Strutt's Teagle.

The operator caused the car to ascend by pulling the handle, which released the brake and engaged the clutch. Upward travel continued as long as the operator maintained pressure on the handle. The clutch disengaged and the brake was applied when the operator released the handle. To descend, the operator applied an intermediate amount of pressure on the handle, releasing the brake, and the car would descend, its speed regulated by the brake.

The innovation in Waterman's elevator was that it was controlled from within the car by means of what we would call a joystick, rather than the bothersome shipper rope that is prohibited today.

By 1850, George H. Fox and Co., a Boston firm, was building freight elevators that were safer and more efficient. Fox replaced meshing spur gears with a worm gear attached to the winding shaft. This arrangement is superior because it is self-locking. The worm can turn the gear, but the gear cannot turn the worm. Consequently, a separate brake was not required for the hoist, which would hold its position when the driving belt was disengaged. This arrangement meant less chance of a car and occupants falling to the bottom of the elevator shaft due to mechanical failure in the drive system.

Safety was further enhanced by other innovations by Fox and Co. One was the replacement in 1852 of traditional hemp rope by stronger and more wear-resistant steel wire rope. The other innovation was a safety brake, which could stop the car from free falling in the event of rope failure. This brake, however, was not automatic and depended upon quick action by an alert operator.

Falling cars were still a severe hazard, but after 1850 new developments in elevator technology greatly reduced the number of occurrences.

In the mid-nineteenth century, William Adams and Co. manufactured freight elevators in Boston. In 1859, one of their freight platforms in a group installation dropped to the bottom of its shaft. An engineer for the firm, inspecting the damage, found that it was not as severe as might be anticipated. He concluded that the hoistway, as built, happened to be relatively airtight, and as a result, the air as it was compressed below the falling platform acted as a cushion and slowed its fall. This suggested a way to mitigate these disasters, and in fact the idea was patented and hardware developed and marketed.

Elisha Graves Otis Invents Safeties

Another very active key figure in the evolving elevator industry in mid-nineteenth century America was Elisha Graves Otis (1811–1861). His Improved Elevator of 1854 incorporated an automatic safety mechanism, which in the event of rope failure as shown in Figure 1-3, would activate automatically.

All elevators, of course, had guide rails, which were necessary to prevent the suspended car from swinging from side to side, striking the hoistway walls. The Otis Improved Elevator was a variation on existing rack-and-pinion drives, in which the rack was attached to the guide rails. In the new design, the teeth curved upward rather than extending perpendicular from the rack. The brake, relocated below the cross beam at the top of the platform or car, consisted of safety dogs connected to a spring and the hoisting rope. Because the rope, as long as it remained intact, supported the freight platform or passenger car, the spring remained compressed and held the safety dogs away from the rack and the elevator functioned as expected. In the event of a break in the rope or if for any reason it lost tension, the safety dogs would engage the upward angled rack teeth, preventing the car or platform from falling.

Otis was an accomplished mechanic and very inventive builder of elevators, always sensitive to safety issues. However, on the financial side his business failed to prosper despite the success of his Improved Elevator with its advanced safety mechanism. Beginning around 1860, nearly all traction elevators incorporated his braking system in one form or another.

FIGURE 1-3 Elisha Graves Otis cuts the hoisting rope, demonstrating his new safeties, still used, that prevent an elevator car from free falling. (*Wikipedia*)

Just three months after receiving his patent, Elisha Otis died of natural causes. His business flourished under the ownership of his sons, who reconstituted the firm as N.P. Otis and Brother. The company prospered under the inventive and financial skills of Norton and Charles Otis. They quickly adapted to the new post-Civil War environment, in which the focus now included passenger elevators built for the new generation of higher-rise hotels, shops, and office buildings.

Hydraulic Elevators

At about the same time that these developments in traction elevator safety and reliability were occurring, in England and continental Europe as well as in the U.S., hydraulic elevators were emerging in low-rise applications. Here we are talking about water pistons, as opposed to the hydraulic oil machines of today. Typically, the water supply was from a high-capacity pump system or reservoir. The water pressure would cause the car to rise to the top floor or as high as required. Then, a discharge valve permitted the car to descend at a measured pace due to its own weight.

Hydraulic elevators had some intrinsic advantages in low-rise applications. Those running off a natural or impounded reservoir had no further fuel costs, and unlike steam power, there were not the tasks of moving in coal and disposing of ashes. They were simple and quiet. In the event of piston failure, the car or platform would not free fall, its speed of descent regulated by the size of the rupture.

Bedrock or a high water table could make for a difficult installation. Builders of hydraulic elevators could then, however, resort to hybrid designs, standing the cylinders vertically above grade outside the buildings or laying them down horizontally. These installations required additional wire rope and pulley mechanisms, compromising the advantages of simplicity and safety.

Just as the nineteenth century was a time in which elevators evolved from primitive lifts to becoming a defining fact in the great cities of America and Europe, so in the ninth decade of that century did the electric motor assume new forms, enabling it to replace coal-burning steam power.

Throughout the 1870s, hydraulic (water) elevators were installed in great numbers. Drive configurations and structural variations proliferated as did the number of manufacturers building them. Additionally, there were many exclusively wire-rope machines being built and installed, with great innovations that made them safer and more efficient. Still, steam power, which was noisy, hot, and required frequent human intervention, powered most elevator installations.

Edison and Westinghouse

Then, beginning around 1880, the DC electric motor changed everything.

The first electrical distribution system was Thomas Edison's 110-volt DC utility in lower Manhattan, intended for indoor residential and commercial use. It was energized in 1882, followed four years later when George Westinghouse began building an AC system, enabling the use of transformers to increase the voltage for efficient transmission and lower it for users. AC eventually eclipsed DC, but meanwhile Edison commenced large-scale DC motor production and for many decades these motors remained in use in many applications for which they were better suited than AC motors, notably in elevators.

DC motors could be run off an AC power supply by means of a simple motor-generator set, often in a single enclosure with no exterior shafts, and later by tube-type and inexpensive solid-state diode rectification. The reason a DC motor was at the time preferable to an AC motor was that, although both could be reversed, the speed of an AC motor could not be easily varied, as required to operate an elevator. In contrast, DC motor speed is varied simply by adjusting the voltage applied to the armature or current applied to the field circuit.

Nikola Tesla, working with George Westinghouse, developed three-phase AC power distribution and he invented the highly efficient and maintenance free three-phase induction motor, shown in Figure 1-4, which quickly permeated industrial facilities worldwide. But since it was essentially a single speed device, it was not suitable for elevator power until the 1960s, when the variable frequency drive (VFD) was introduced. This consisted of electronic circuitry that permitted users to run AC induction motors at lower (or even higher) than rated speed by means of pulse-width modulation (PWM), which we will discuss in detail in Chapter 3.

FIGURE 1-4   Tesla's AC induction motor was not suitable for elevator use until the 1960s, when the variable frequency drive made speed control possible. (*Wikipedia*)

When electric motors were first suggested for elevator power, the public was skeptical. There had been a number of power line fatalities as new distribution systems were being constructed, and fire hazard was perceived to be an issue in high-rise buildings compromised by wooden hoistways piercing multiple floors. Early electric codes such as NEC, first issued in 1897, decisively confronted these hazards, and soon electric motors became part of everyday life.

The first elevator motors powered building-wide belt driving shafts in manufacturing facilities, so they were external to the elevators. But space and manufacturing costs could be saved by integrating the motor directly into the elevator assembly. That was accomplished before 1890, and is how it remains today.

Frank Sprague

Before the end of the nineteenth century and continuing to the present, electric elevators improved, with new designs becoming safer and more efficient. A key figure in this development was Frank Sprague, shown in Figure 1-5.

Electric motors and their applications in human transportation were Frank Sprague's life. After graduating from the U.S. Naval Academy and a short stint on ship and in Europe, the young electrical engineer joined Edison's large assembly of electricians, mechanics, and glassblowers in the lab at Menlo Park. While Edison was focused on producing a practical electric light bulb, Sprague wanted to develop a DC motor that would maintain RPM under varying loads. Edison was temperamental, but went along with this idea.

FIGURE 1-5   Frank Sprague (1857–1934) (*Wikipedia*)

Prior to 1880, electric motors were repurposed electric generators, then known as dynamos, which had preceded them. It had been found that voltage applied to what had been the generators' output terminals would cause them to turn. These devices would actually run and could be configured to perform work, but they left a lot to be desired.

Sprague had some big ideas. He envisioned a DC motor that could run a loom, hoist, pump, blower, or machine tool. His highest ambition, eventually realized, was to build powerful motors that were reliable and capable of powering railroads, replacing the inefficient, smoky, and dangerous steam engines of the day.

Dynamos repurposed as motors bogged down under heavy load, and while this didn't make much difference in some applications, in others these primitive devices were not suitable. A skilled mechanic was needed during running hours to advance or retard the brushes and adjust field strength for various loads and RPMs.

Sprague, at this juncture and throughout his life, demonstrated that Edison wasn't the only electrical and mechanical genius. While Sprague has had less impact than Edison in the popular imagination, in many ways he was more advanced and insightful. Sprague built an electrical motor that maintained constant speed under varying load. Rather than the steam engine's mechanical governor, Sprague's electrical motor incorporated a reverse winding that automatically varied field strength in response to speed and loading. He solved the problem of brush position not by moving them physically, but by rotating the magnetic field to achieve the required alignment.

Since Sprague was working for Edison, the improved motor design at this point belonged to Edison. Sprague evidently saw the writing on the wall, and shortly thereafter tendered his resignation.

While Edison continued to refine his incandescent light bulbs and DC power generation and distribution system, Sprague, after eleven months working in Edison's large organization, formed the Sprague Electric Railway and Motor Company. His lifelong project was to move rail traffic by means of electric motors. In this he was very successful and was renowned among electricians and transportation workers as "the father of electric traction." He came to define these words to include vertical as well as horizontal traction. His elevator work was a relatively brief interlude, but its impact was enormous. After completing some difficult early streetcar and railway projects, he turned his attention in 1889 to elevator design and construction.

Sprague, together with his old friend Ed Johnson and elevator manufacturer Charles Pratt, rented a factory building and in 1892 formed Sprague Electric Elevator Company. Ed Johnson was the legal and financial specialist. Sprague provided electrical expertise, and Pratt was the mechanical engineer. Together they planned to offer two very different types of elevators. For low-rise buildings, a conventional drum-type elevator would be reconfigured with a reversible, adjustable-speed electric motor replacing the steam engine.

For high-rise applications, a faster machine would consist of a large, threaded-steel shaft placed horizontally and powered through a gearbox by an electric motor. A large nut would move along the turning shaft, driving a cable pulley. The contraption actually worked, and in fact dominated the industry until shortly before the turn of the century.

After constructing a small prototype in their new Manhattan facility, the firm secured a contract to install a similar elevator in the Grand Hotel in New york City.

There were problems in this installation. The control system, which had been satisfactory in Sprague's electric trolleys, did not provide the smooth performance required in an elevator.

Sprague's electrical expertise was severely challenged. First, he built an improved resistance network, known as the grid, for the controller. This smoothed out the elevator motion, but the resistance network and controller mechanism heated and contacts had to be replaced.

The elevator was put back in operation, but after a few days at an upper floor the ascending car suddenly dropped, its speed doubling, coming to a stop after striking the bumpers at the bottom level.

Fortunately, there were no injuries. The cause was determined to be a defective motor that ran the reversing lever in the controller. The sudden reversal damaged the safeties, permitting the car to drop.

Soon redesigned safeties and controls were in place and the elevator resumed normal operation. This did not solve the network problem. Eventually, the firm built and installed a new controller with heavier contacts, but the problem persisted. Sprague favored a cast iron grid, which turned out to work on a long-term basis. The Grand Hotel signed off on the project and Sprague Electric Elevator Company moved on to another project, the Postal Telegraph Building. This was to be located close by on Broadway and be far bigger and faster than the Grand Hotel installation. There would be four local and two express elevators, rising 14 stories above street level.

The immense screw and traveling nut mechanisms resulted in heavy loading and increased friction, which Sprague intended to mitigate by incorporating captive steel balls within the nuts.

After mishaps and delays, the Postal Telegraph Building installation performed flawlessly in tests and was placed in operation. It was a prestigious project, and the Postal Telegraph Company Building Committee was well satisfied. But unfortunately, the country was enduring a severe financial depression. New building had halted, and orders were not coming in. Sprague, never one to let up, had some ideas for improving elevator safety and efficiency, and the business slowdown allowed time to work on them. One innovation was the self-centering "dead man's control," which stopped the car if for any reason the control was released by the operator. Another innovation was an automatic elevator that incorporated door interlocks and floor alignment. Sprague stuck with the screw and nut design, rather than going with an improved traction drive, which together with his electric motor became the wave of the future. The business climate improved and by 1895 new orders soared and the company moved to a larger facility across the Hudson River, in New Jersey.

In 1898, Sprague decided to return to his first love, railroad electrification. He sold the elevator business to Otis for $1,000,000, retaining royalty rights to two-thirds of foreign business and rights to lease back plant and equipment for five years. With that transaction Sprague became much less of a presence in the elevator world, in which Otis was now the most prominent player.

The Otis Story

Elisha Graves Otis (1811-1861), a skilled builder and mechanic, in 1850 found his way to Albany, New york, where he was employed to manage a bedstead factory. Relocating in nearby Bergen, New Jersey, in 1851 and then in yonkers, New york, he organized and managed successive bedstead facilities, and in the yonkers factory he built a freight elevator. Soon, he established an independent company, which by 1853 was building and installing freight hoists in nearby manufacturing facilities.

In early 1854, as we have seen, one of the defining events in elevator history occurred. At an exhibition in New york's Crystal Palace, Otis ascended in one of his fully-loaded open freight hoists, and in the presence of astonished onlookers, cut the hoisting rope. Rather than crashing to the floor, a frequent cause of fatal elevator accidents, the platform dropped a short distance and then stopped.

Otis had demonstrated the effectiveness of his great invention. Safeties, as they were (and still are) called, in response to breakage or loss of tension in the rope, automatically, which is to say without human intervention, gripped racks attached to guides, bringing the car or platform to a stop. These safeties, in one form or another, were universally adopted in the elevator industry and have saved many lives.

Otis, in 1855, established the Union Elevator Works. The firm sold a gradually increasing number of hoists in the years that followed. They were powered by water or steam engines, which were optionally furnished, and always with the new safeties.

Elisha Otis was a great mechanic and inventor, if not always successful financially. Following his death in 1861, his sons, Norton and Charles Otis, took over the firm, renaming it N.P. Otis and Brother. As inventors and builders, their skills equaled Elisha's, and financially they succeeded where their father had been challenged.

After the American Civil War, in 1867, Norton and Charles again renamed the firm Otis Brothers and Company. In the years that followed, the organization, by means of intense research and development and aggressive marketing, became the preeminent powerhouse that it is today. By the end of the nineteenth century, through a series of mergers, stock acquisitions, and purchases, Otis absorbed its major competitors.

Electrification

First Edison, and then Tesla and Westinghouse had built electrical distribution systems capable of supplying power where needed. Clearly, the electric motor was the wave of the future. At first, elevators moved from steam engine to electric power merely by substituting an electric motor for the connection to mill shafting or the steam engine. This arrangement worked reasonably well, but Otis Brothers and

Company between 1887 and 1889 realized further advantages in fully integrating the motor into the elevator mechanism. This development occurred at a time when new, taller buildings were proliferating in New york City and other urban areas. Otis began selling passenger elevators for the new generation of hotel and office buildings. An operator in the car controlled a reversible, multispeed DC motor from inside the cab, at first by means of a shipper rope and later a dead man's rheostat. A separate worker tending a steam engine was no longer required.

In 1892, Otis Electric Co., jointly owned by Otis Brothers and General Electric, was created for the purpose of designing and building elevator-specific electric motors for Otis Brothers and Company.

In the years that followed, Otis Brothers continued to acquire competing elevator manufacturers, and in 1898 they created a gigantic umbrella entity known as Otis Elevator Co., which is now well into its second century of operation.

From 1900 to the present, there have been numerous refinements in elevator technology, some incremental, some revolutionary. The trend has been to make vertical transportation safer, faster, more efficient, and more reliable. Three principle developments that revolutionized elevators in the twentieth century are:

- The automatic elevator
- The low-rise hydraulic elevator
- The VFD, which enables use of AC induction motors to power elevators

We will discuss each of these.

Automatic Elevators

At first elevators were operated by means of clumsy and not always reliable shipper ropes that passed through holes in the car floor and ceiling. (They are prohibited in the current ASME A17 Safety Code for Existing Elevators and Escalators.) Car and hoistway doors had to be opened and closed manually, and until the development of the door interlock, a door could be left open, sometimes resulting in fatal accidents.

The first automatic door mechanism was built and patented in 1887 by Alexander Miles, an African-American inventor in Duluth, Minnesota. The door opens and closes by means of a series of rollers and levers. After a car has stopped at a floor, a flexible belt extending the length of the shaft opens the shaft door. The car door also opens automatically. Both doors close before the car proceeds to the next stop.

Still, a human operator started and stopped the car by means of a controller inside the car, as shown in Figure 1-6.

FIGURE 1-6   Otis manual elevator controller. If for any reason the operator released the handle, it would return to the neutral position and the car would stop. (*Wikipedia*)

Automatic Motion Controllers

Rudimentary elevator automation first appeared in the 1920s. At the time there were no microprocessors or solid-state components, but digital logic could be accomplished by means of mechanical relays. These were cumbersome by today's standards, and had some disadvantages. They were slower-acting, consumed more energy, were less reliable, and more costly. Still, they worked surprisingly well in elevator applications, and in contrast to today's computer-controlled devices, there were never system-wide crashes.

The first relay version was complex and had few of the features we expect in fully automatic elevators. It was known as the selector. It had numerous mechanical parts including a magnetic tape attached to the top of the car. This tape, as the car traveled, caused mechanical gears to move in response. The gears controlled speed, position, and door operation. A human operator was still required, but car leveling and stopping were simplified.

In 1924 Otis introduced Signal Control, which was a fully automatic elevator system, still with mechanical relays. Throughout the 1940s and 50s, other manufacturers introduced enhanced relay-controlled automation, permitting the car to bypass floors when fully loaded.

Microprocessor-based controllers were introduced by Otis in 1979. The Elevonic 101 was a true motion controller, overseeing all aspects of elevator operation. Another Otis product, Elevonic 401, offered in 1981, was fully computerized.

All-in-one, microprocessor-based controllers, which are a product of China, are currently widely used. They are compact and consume far less energy than previous motion controllers. These units sense car position and door status and are capable of managing large group installations, with human intervention necessary only rarely in the event of sensor, termination, or wiring failure.

Software as a Service (SaaS) enables remote monitoring of group installations via web browser, and it will alert technicians and building managers of imminent or actual malfunctions. Remote monitoring systems are offered by Otis (REM), ThyssenKrupp (Vista), Schindler (Servitel), Kone (KRM) and Mitsubishi (ELE-FIRST).

We will discuss the inner workings of the contemporary motion controller in Chapter 9.

Hydraulic Elevators

Speaking now of hydraulic elevators, we are no longer concerned with the water-powered affairs that were prominent in the late nineteenth century. The newer hydraulic elevators had some important differences. For one thing, the fluid that characterized them was not water, discarded after each cycle, but hydraulic oil, which, with an anti-foaming additive, resembles a lightweight non-detergent motor oil. It is never discharged into the ground, a water body, or into a waste disposal system, but instead is reused until with multiple heatings, it eventually breaks down and must be changed like automatic transmission fluid in a motor vehicle. When the hydraulic piston is retracted, the oil returns to a steel tank located in the machine room, a reservoir where it cools. When the hydraulic piston is fully extended, enough oil remains in the reservoir so that the submersible hydraulic pump is covered.

We'll have a lot to say about different hydraulic elevator configurations in Chapter 2, Types of Elevators, and about troubleshooting them in Chapter 5, Troubleshooting Elevator Systems.

Variable-Frequency Drives

As pointed out earlier in this chapter, Nikola Tesla's brilliant invention of the AC induction motor in conjunction with three-phase power was not much help for the elevator until the development of the VFD in the 1960s.

Also called adjustable-frequency drive, variable-voltage/variable-frequency drive, variable-speed drive, AC drive, micro drive, and inverter drive, the VFD was developed in response to the need to enable the very efficient, reliable, and inexpensive AC induction motor to run at infinitely variable speed and torque levels without wasteful heat accompanying rheostat-controlled voltage as in the DC motor.

To run a DC motor off the usual AC power supply required a motor-generator set or diode rectifier. This was not as great a problem as one might think. After all, a VFD requires high-power DC for the solid-state inverters in the output stage, so this rectification is provided in the front end. Virtually all elevator motors, AC or DC, use full-power rectification somewhere along the line. But the DC motor was more expensive to manufacture and the brushes and commutator required regular maintenance.

There were some early relatively crude VFDs, such as the rotary machines patented by General Electric in the early twentieth century, but they were not generally used in elevator applications. VFD technology improved in stages over the years. Before 1958 there had been various mechanical systems, but VFDs were not widely used until the introduction of silicon-controlled rectifiers (SCRs), which set the stage for subsequent improvements. In the early 1960s, the cost of SCRs dropped and VFDs became available for manufacturing applications. After the late 1960s, analog control circuitry with digital input was introduced. With phase-locked loops for synchronization, motor drives became less subject to noise and hence more reliable.

Throughout the 1970s, large-scale integration (LSI) improved VFD reliability, and cost further declined. In the late 1980s, bipolar pulse-width modulation (PWM) came on the scene and switching frequency increased. Insulated gate bipolar transistors (IGBTs) emerged.

Today, VFD three-phase (and for small applications, single-phase) drives are widely used wherever AC motor speed control is required. VFDs are marketed with dedicated AC induction motors, or off-the-shelf motors with suitable bearings and cooling may be obtained.

VFDs will be covered in more detail in Chapter 3, AC and DC Electric Motor Maintenance, VFD Troubleshooting, and Diagnostic Procedures, Chapter 4, Advanced Motor Repair, and Chapter 5, Troubleshooting Elevator Systems.

Regenerative Elevator Drives

Regenerative braking is a valuable energy recovery strategy. It has been used by railroads in long downhill grades, and it is a logical solution for elevators, which do a lot of braking.

Electric motors operate as or can be configured to operate as generators. In this mode, energy that would otherwise have to be dissipated as heat is fed back into the building to power other loads, and/or returned via reverse metering to the utility where it is credited to the customer's account.

Virtually all traction elevators have counterweights, so energy consumption is divided equally between upward trips when the car is heavily loaded and downward trips when the car is lightly loaded. Actually, regenerative braking was used over a hundred years ago in elevators as well as cranes, which also work in a downward-going mode.

Regenerative braking benefits the building owner by reducing utility bills, and it benefits our planet by reducing carbon emissions inherent in electrical generation that relies on fossil fuels.

Another benefit is that in hot weather the heat generated by purely dynamic braking does not have to be offset by motorized fans or by the building's air-conditioning system.

There are other energy-saving strategies that have emerged. Replacing relays, solid-state controls dissipate less heat. Sensors in conjunction with software cause the elevator to enter a sleep mode, temporarily switching off in-car lighting and ventilation when not in use. Car stops are batched, reducing waiting time. Double-deck cars, one above the other, stop at even- and odd-numbered floors simultaneously, saving energy and reducing the size of group installations. LED lighting cuts energy costs dramatically, and it can be retrofitted without even changing the fixtures.

Recent Developments

Two important contemporary concerns in elevator technology are reducing energy consumption and dealing with ever-greater heights in the latest high-rise buildings.

Buildings consume about 40 percent of the world's energy, and of this, elevators require between two and ten percent. Obviously, if ways can be found to reduce elevator energy consumption, building owners and indirectly individual tenants will realize capital savings. And more important, significant progress will be achieved in reversing climate change.

The shift from DC to VFD AC induction motors, regenerative braking, better software, more efficient cable, and counterweight systems and LED lighting are examples of energy-saving measures already in place, though not fully realized.

Hydraulic elevators offer advantages in low-rise applications, but for anything over five stories, traction elevators with the exception of some new alternate designs continue to be the focus. In cities and in widening circles around them, enormous high-rise projects are causing us to rethink traction elevator design.

First we need to consider, in the context of energy efficiency, geared versus gearless motors. In geared elevators, the motor drives a gearbox, which turns the sheave at a substantially lower speed than the input shaft RPM. In gearless elevators, the motor turns the sheave directly, eliminating gearbox loss in heating the oil. Gearless drives reduce energy loss substantially, and moreover gearless motors, substituting torque for RPM, have a longer service life. The initial cost is greater, but long-term savings are substantial.

Another area of concern, particularly in taller buildings, is elevator rope. Longer steel rope means more weight, sometimes to a point where the rope has difficulty holding its own weight. In the tallest buildings, wire rope may weigh several tons, comprising 70 percent of the total load.

Elevator manufacturers have been confronting this problem by developing lighter-weight alternatives. For example, Schindler has introduced Aramid fiber rope, which is both stronger and lighter. Gen2, offered by Otis, consists of thin cables enclosed within a polyurethane outer sheath. Mitsubishi has introduced a stronger, denser rope composed of steel wire arranged in concentric layers. Kone's UltraRope consists of a carbon-fiber core. The rope is only ten percent the weight of conventional wire rope.

Since these ropes are stronger and lighter with smaller profile, power requirements are reduced.

Another very significant innovation, by ThyssenKrupp, is the Twin System. Two cars travel independently in a single shaft. Besides moving passengers more efficiently, the number of shafts is reduced, saving space on each floor in a group installation, which can now be rented.

Space Elevator

Elevators, beginning in the nineteenth century and continuing to the present, have evolved from primitive platforms that lifted freight in factories and mines, to powerful multi-car systems that convey materials and human passengers to offices and residences approaching a mile above street level. There is every reason to believe that they will continue to grow. A space elevator, shown in Figure 1-7, has even been envisioned. It would transport freight and humans from the surface of the earth to geocentric orbit and beyond.

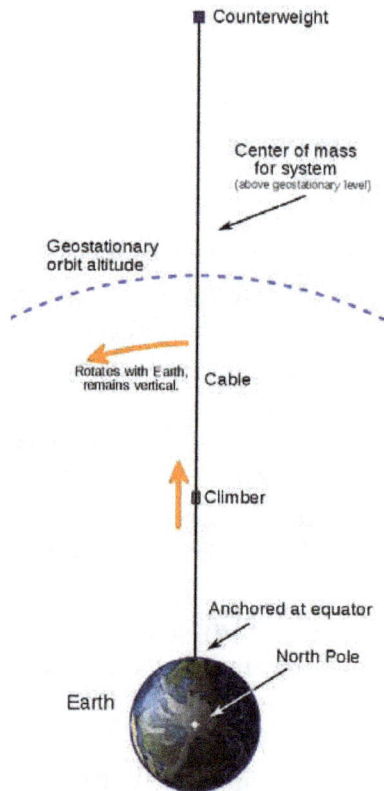

■ Counterweight

Center of mass
for system
(above geostationary level)

Geostationary
orbit altitude

Rotates with Earth,
remains vertical. Cable

■ Climber

Anchored at equator

North Pole

Earth

FIGURE 1-7 Idealized diagram of a space elevator, not to scale. (*Wikipedia*)

A cable or tether would be anchored to the earth and extend into space. Vehicles would climb this cable, lifting passengers and freight to orbiting space stations without benefit of huge shuttles. The cable would extend well beyond the system's center of mass, to a heavy counterweight, which due to its position well beyond the geostationary orbit, would exert a powerful counter-gravitational force, pulling the cable tight.

A continuous stream of climbers (not human, but mechanical) could ascend the cable, passing others returning in the opposite direction to earth. Escape velocity would not have to be attained. It would be a relatively slow journey, requiring about a week. Konstantin Tsiolkovsky conceived the space elevator in 1895, near the end of the century that defined elevator technology. Rather than a cable under tension because it was hanging from the counterweight, it was a tower, sitting on the ground and subject to great compressive force.

There are some major difficulties in building the space elevator, primarily having to do with fabricating the cable, which has to support its own enormous weight without breaking. As of now, no known material can do that. But carbon nanotubes should work. Development costs are high, but once built, the space elevator will transport passengers and freight quite economically. Space elevators will eventually be constructed on other planets, moons, and asteroids. For these lighter bodies, the challenges are less formidable but the rewards are not as great.

Lighter materials such as Kevlar would be suitable for constructing extra-terrestrial cables. In 1975, Jerome Pearson introduced the idea of a tapered cable. Maximum tension on a space elevator cable would be at geosynchronous altitude, so the cable would have to be thickest there and taper carefully as it approaches earth. The concept of a space elevator became more realistic after the development in 1990 of carbon nanotubes. In 2000, Bradley Edwards proposed a flat ribbon rather than the round cable of previous designs, because it would be less vulnerable to damage from meteors and space debris. Moreover, climbers could use rollers to travel upward. Since then, numerous feasibility studies have concluded that the space elevator is a valid concept, and it will profoundly affect human history as we continue on our greatest journey.

STUDy QUESTIONS

1. Who made the first extant reference to an elevator?
   A. Archimedes
   B. Vitruvius
   C. Plato
   D. Aristotle

2. Early nineteenth century elevators:
   A. were powered by work horses
   B. were steam-powered
   C. were powered by electric motors
   D. rose to a height of 20 stories

3. William Strutt's elevator:
   A. was in a coal mine
   B. had no pulleys
   C. ran off a flat belt
   D. carried passengers and freight

4. Hotels and office buildings used elevators beginning around:
   A. 1875
   B. 1865
   C. 1855
   D. 1845

5. Henry Waterman's elevator in Manhattan:
   A. did not require an outside attendant
   B. was powered by an electric motor
   C. was a hydraulic elevator
   D. had a clutch that disengaged when the operator released the handle

6. In George Fox and Co.'s freight elevators:
   A. there were frequent mechanical failures
   B. meshing spur gears with a worm gear became obsolete
   C. a separate brake for the hoist was required
   D. wire rope replaced hemp rope

7. Elisha Graves Otis:
   A. was enormously successful financially
   B. pioneered the use of safeties
   C. specialized in hydraulic elevators
   D. built traction engines throughout the United States

8. In a hydraulic elevator installation:
   A. if bedrock was encountered, a hybrid design was needed
   B. cylinders could be installed vertically only
   C. noise was a severe problem
   D. complex rope and pulley assemblies were required

9. Nineteenth century hydraulic elevators:
   A. used no coal
   B. used oil for hydraulic fluid
   C. rose to unprecedented heights
   D. would free fall if the piston failed

10. Electric motors replaced steam in elevators:
   A. before the American Civil War
   B. in the 1920s
   C. after 1900
   D. beginning around 1880

For answers, go to Appendix A.

TYPES OF ELEVATORS

Virtually all modern elevators fall into one of three categories, with some exceptions, variations, and models that combine elements of two or even all three types. These types are traction elevators, hydraulic elevators, and machine room-less elevators.

Traction Elevators

Traction elevators are the most common type. They are characterized by multi-strand steel "ropes" that in a typical design are attached to a hitch plate at the top of the car. There may be six or more of these cables, each capable of lifting the car and occupants. The cables pass over and are driven by a deeply slotted sheave, two or more feet in diameter, at the top of the shaft, with the other ends attached to a counterweight. (In one design, two cars move synchronously in opposite directions, each functioning as the other's counterweight.)

Instead of traditional ropes, some manufacturers, notably Otis, Schindler, and Kone, have introduced very light steel belts, with carbon fiber cores and high-grip coating. Lubricant is not necessary, and energy consumption is reduced, especially in high-rise applications.

Traction elevators, shown in Figure 2-1, may be geared or gearless. In the geared design, a higher-speed electric motor is coupled to the hoisting sheave by means of a worm-and-gear speed reduction unit, which turns the hoisting sheave. This arrangement has the advantage of requiring a smaller motor. The car travels at speeds of 125 to 500 feet per minute, with lifting capacity of up to 30,000 pounds. An electric brake stops the car as required and holds it at floor level.

The gearless traction elevator design permits car speeds in excess of 500 feet per minute. The counterweight, sized to equal the weight of the car plus half the weight of a car full of passengers, reduces the load on the motor. Car speed is a function of motor RPM and sheave size, typically two to four feet in diameter. To achieve proper car speed, this huge sheave turns at 50 to 200 RPM.

FIGURE 2-1   Traction elevator motor and drive. (*Judith Howcroft*)

Safeties

In addition to the multiple lifting cables shown in Figure 2-2, safety is provided by car brakes that are engaged if the car were to begin falling at greater than a specified speed or if for any reason tension is lost on the hoisting ropes. A clamp closes on the steel governor cable and this causes brakes to engage the guide rails, stopping the car not too abruptly, but quickly enough so that it does not gain excessive momentum.

If the maximum vertical travel is greater than 100 feet, a system known as compensation is used. This consists of an additional set of cables or a steel chain, one end of which is attached to the bottom of the car and the other end is attached to the bottom of the counterweight. As the car rises, more of this chain or cable is lifted to balance the shorter amount of hoist rope between the car and sheave, and as it descends and the counterweight rises, this weight is added to the counterweight. This equalizes the amount of work required of the motor.

When compensation cables are used, an additional sheave at the bottom of the shaft keeps them in place. If they take the form of chains, they are guided by a horizontal bar.

FIGURE 2-2 In a traction elevator, steel ropes engage deep grooves in the sheaves. (*Renown Electric*)

In traction elevators there are several possible roping configurations. Suspension ropes attach to a hitch plate above the car, or they are underslung below, loop over the sheave, and pass down to the counterweight. Depending on the size of the car, there may be as many as eight, sometimes more, of these hoisting ropes, typically ⅝ inch diameter.

Low-speed elevators with geared motors generally have a single-wrap arrangement, where the rope passes over the sheave once and connects to the counterweight. Double-wrap is used with higher speed elevators having gearless motors.

1:1 roping consists of rope that is connected to the counterweight and travels as far as the car travels, in the opposite direction. It is used with geared traction systems and high-speed elevators. 2:1 roping consists of a sheave attached to the top of the counterweight. The rope moves twice as far as the car. This configuration is used on machine room-less, bottom-drive traction, gearless traction, and freight elevators. When higher capacity is required, 4:1 roping is used. The rope moves four times as far as the car.

In all cases guide rails are necessary. Otherwise the car, suspended at the end of the cable, would swing from side to side, hitting the walls of the shaft. In a new installation or making changes, wiring (in conduit), junction boxes and the like can be mounted on the shaft walls. It is important that they do not protrude beyond the guide rails, so they are not in the path of the car. There may be very little clearance in this area.

In today's world, saving energy is a focus in building services. In an elevator installation, a regenerative drive accomplishes this by using the electric motor as a generator to return electrical energy to the facility and/or utility when a full elevator, which is heavier than the counterweight, descends or when an empty elevator ascends.

A traction elevator consists of steel ropes that raise and lower the car at a measured rate. The steel ropes lie in grooves milled into the sheave. These grooves serve two purposes—they keep the ropes separate and in place so that they don't bunch up and tangle, and they provide much greater traction than if the ropes were to wrap about an ungrooved cylinder, in which case there would be only a single line of contact.

Regrooving

Naturally, with all the fast starts and stops, heavy loading, and continuous use, the steel ropes have to be replaced and the sheaves need to be regrooved periodically. How often depends upon various factors. In a group installation, you may be able to take one car at a time out of service in order to have the regrooving done and/or ropes changed. If the building has only a single car, you'll need to do some careful planning and scheduling. For maximum rope and sheave life, it is important that all ropes have equal tension. Ropes should be checked frequently for signs of wear. All ropes should protrude an equal amount from their respective grooves. Worn sheaves cause premature wear on ropes, which then accelerates sheave wear, creating a mutual destruction scenario. Frequent inspection and maintenance as required are essential. An advantage of the traction elevator is its very safe and efficient braking system. The brake, as in automobiles, may consist of a large drum with brake shoes or a smaller disc with pads. The brake assembly is located close to the motor and/or gearbox. It is electrically actuated. Power is required to keep the brake disengaged, and when power is interrupted, the brake is applied. Thus, in a power outage, the brake is applied so the car cannot move.

Power is applied and interrupted simultaneously at the motor and at the brake. Power is interrupted in an outage when the motion controller senses a serious fault or shuts down for some other reason, when a human shuts off the disconnect, or when the car is intended to stop at a landing. If the motor were to stop without the brake being applied, the net weight of the car and counterweight would probably, depending on the gearing and loading, cause the motor to spin, allowing the car to move. Conversely, if power were to be interrupted at the brake but not the motor, the brake would rapidly heat with all that would entail.

Hydraulic Elevator

In limited applications, the simplicity and user-friendly nature of hydraulic elevators makes them the preferred choice.

The most common installation, shown in Figure 2-3, involves a large hydraulic piston that extends underground beneath the building to a depth equal to the amount of car travel, measured at the bottom of the car floor. In practice, this type of installation is limited to not much over four floors, so it is never seen in high-rise buildings. That being said, they are often preferred in low-rise buildings where there is no bedrock. There are boring techniques that permit a hydraulic cylinder to be installed in a retrofit situation under an existing building.

An underground oil leak, particularly where there is a nearby aquifer, is a major environmental catastrophe. For this reason, some manufacturers have discontinued hydraulic elevators, while others have developed high-performance cylinder liners to contain any oil that could otherwise escape.

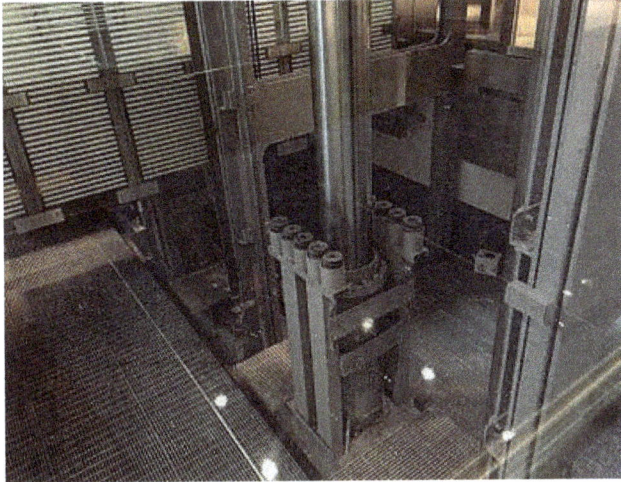

FIGURE 2-3 A below-grade hydraulic piston drives the car to a maximum height of about four stories.

In low-rise applications, hydraulic elevators are widely used and there are a great many currently in service. Some have above-ground cylinders, which solves a few problems but gives rise to others, such as dedicated space that is required, and greater complexity due to the nature of the hybrid traction and hydraulic design. Some hydraulic designs have telescoping pistons, which reduce the amount of excavation and permit greater vertical travel, at the expense of greater complexity. An advantage in the conventional single-cylinder, below-grade design is an intrinsic added degree of safety because in the event of rupture, the car can fall no faster than permitted by the escaping oil.

In a typical installation, submersible motor/pump units are bolted together and located in a 50- to 100-gallon oil reservoir, which is located in the machine room adjacent to the motion controller. The small motors are very reliable and long-lasting, because immersed in oil there is excellent heat dissipation and the windings, encapsulated in epoxy, are exceptionally well insulated and protected from grounding out.

There are a number of conditions that will cause an elevator to cease operation. One of these, in the hydraulic elevator, is elevated oil temperature in the reservoir. High oil temperature can have many causes, including among others aged and inefficient oil, low oil level (a leak!), inefficient pump operation, continuous operation of the elevator especially with heavy loading, high ambient temperature in the machine room, and so on. All these factors work together. you can have some idea where you stand by attaching a thermometer to the outside of the tank. The top of the tank should never be used as a catch-all, especially for cloth or similar objects that could impede heat dissipation. The machine room should have adequate ventilation, and if this depends upon a fan, failure of the fan motor could be an issue.

ASME A17.3, Safety Code for Existing Elevators and Escalators, Part IV, contains design and installation requirements specifically relating to hydraulic elevators. It opens with a declaration of scope, stating that Part IV is applicable to both direct plunger and roped-hydraulic elevators. Section 4-1 notes that hoistways, hoisting enclosures, and related construction are to comply with ASME A17.3, Part II.

Section 4.2, Mechanical Equipment, contains four provisions:

- 4.2.1 states that buffers and bumpers are to be provided. Solid bumpers are acceptable instead of spring bumpers where the rated speed is 50 feet per minute or less.
- 4.2.2 states that car frames and platforms shall conform to the requirements of Section 3.3.
- 4.2.3 states that car enclosures are to comply with Section 3.4.
- 4.2.4 states that capacity and loading are to comply with Section 3.7.

Section 4.3 pertains to driving machines:

- 4.3.1, Connection to Driving Machine, states that the driving member of a direct plunger driving machine is to be attached to the car frame or car platform with fastenings of sufficient strength to support that member.

  It is further stated that the connection to the driving machine is to be capable of withstanding without damage any forces resulting from a plunger stop.
- 4.3.2, Plunger Stops, states that plungers are to be provided with solid metal stops and/or other means to prevent the plunger from traveling beyond the limits of the cylinder. Stops are to be designed and constructed to stop the plunger from maximum speed in the up direction under full pressure without damage to the connection to the driving machine, plunger, plunger connection, couplings, plunger joints, cylinder, cylinder connecting couplings, or any other part of the hydraulic system. For rated speeds exceeding 100 feet per minute where a solid metal stop is provided, means other than the normal terminal stopping device (i.e., emergency terminal speed limiting device) are to be provided to retard the car to 100 feet per minute with a retardation not greater than gravity, before striking the stop.
- 4.3.3, Hydraulic Elevators, provides that hydraulic elevators that have any portion of the cylinder buried in the ground and that do not have a double cylinder or a cylinder with a safety bulkhead are to:
  (a) Have the cylinder replaced with a double cylinder or a cylinder with a safety bulkhead protected from corrosion by one or more of the following methods:
    (1) Monitored cathodic protection
    (2) A coating to protect the cylinder from corrosion that will withstand the installation process

(3) A protective plastic casing immune to galvanic or electrolytic action, salt water, and other known underground conditions; or

(b) Be provided with a device meeting the requirements of Section 3.5 or a device arranged to operate in the down direction at an overspeed not exceeding 125 percent of rated speed. The device is to mechanically act to limit the maximum car speed to the buffer striking speed, or to stop the elevator car with rated load with a deceleration not to exceed 32.2 feet per second, and is not to automatically reset. Actuation of the device is to cause power to be removed from the pump motor and control valves until manually reset; or

(c) Have other means acceptable to the authority having jurisdiction to protect against unintended movement of the car as a result of uncontrolled fluid loss.

- Section 4.4, Valves, Supply Piping and Fittings, provides in 4.4.1, Pump Relief Valve:

(a) Pump Relief Valves Required: Each pump or group of pumps is to be equipped with a relief valve corresponding to the following requirements, except as covered by (b):

(1) Type and Location: The relief valve is to be located between the pump and the check valve and is to be of such a type and so installed in the bypass connection that the valve cannot be shut off from the hydraulic system.

(2) Size: The size of the relief valve and bypass is to be sufficient to pass the maximum rated capacity of the pump without raising the pressure more than 50 percent above the working pressure. Two or more relief valves are permitted to obtain the required capacity.

(3) Sealing: Relief valves having exposed pressure adjustments, if used, are to have their means of adjustment sealed after being set to the correct pressure.

(b) Pump Relief Valve Not Required: No relief valve is required for centrifugal pumps driven by induction motors, provided the shutoff, or maximum pressure which the pump can develop, is not greater than 135 percent of the working pressure at the pump.

- 4.4.2, Check Valve, states that a check valve is to be provided and so installed that it will hold the elevator car with rated load at any point when the pump stops or the maintained pressure drops below the minimum operating pressure.

- 4.4.3, Mechanically Controlled Operating Valves, provides that they are not to be used. Existing terminal stopping devices consisting of an automatic stop valve independent of the normal control valve and operated by the movement of the car as it approaches the terminals, where provided, may be retained.

- 4.4.4, Supply Piping and Fittings, states that they are to be in sound condition and secured in place.

- Section 4.5, Tanks, contains in 4.5.1, General Requirements:
  (a) Capacity: All tanks are to be of sufficient capacity to provide for an adequate liquid reserve to prevent the entrance of air or other gas into the system.
  (b) Minimum Liquid Level Indicator: The permissible minimum liquid level is to be clearly indicated.
- 4.5.2, Pressure Tanks, provides:
  (a) Vacuum Relief Valves: Tanks subject to vacuum sufficient to cause collapse are to be provided with one or more vacuum relief valves with openings of sufficient size to prevent collapse of the tank.
  (b) Gage Glasses: Tanks are to be provided with one or more gage glasses attached directly to the tank and equipped to shut off the liquid automatically in case of failure of the glass. The gage glass or glasses are to be located to indicate any level of the liquid between permissible minimum and maximum levels, and are to be equipped with a manual cock at the bottom of the lowest glass.
  (c) Pressure Gage: Tanks are to be provided with a pressure gage that will indicate the pressure correctly to not less than 1.5 times the pressure setting of the relief valve. The gage is to be connected to the tank or water column by pipe and fittings with a stop cock in such a manner that it cannot be shut off from the tank except by the stop cock. The stop cock is to have a T or lever handle set in line with the direction of flow through the valve when open.
  (d) Inspector's Gage Connection: Tanks are to be provided with .25 inch pipe size valve connection for attaching an inspector's pressure gage while the tank is in service.
  (e) Liquid Level Detector: Tanks are to be provided with a means to render the elevator inoperative if for any reason the liquid level in the tanks falls below the permissible minimum.
  (f) Handholes and Manholes: Tanks are to be provided with a means for internal inspection.
  (g) Piping and Fittings for Gages: Piping and fittings for gage glasses, relief valves, and pressure gages are to be of a material that will not be corroded by the liquid used in the tank.
- Section 4.6, Terminal Stopping Devices, provides that terminal stopping devices are to conform to the requirements of 3.9.1.
- Section 4.7, Operating Devices and Control Equipment, provides in 4.7.1, Operating Devices, that operating devices are to conform to the requirements of 3.10.1 and 3.10.2, Top-of-Car Operating Devices, which state that top-of-car operating devices are to be provided and are to conform to the requirements of 3.10.3, except for un-counterweighted elevators having a rise of not more than 15 feet. The bottom normal terminal stopping device is permitted

to be made ineffective while the elevator is under the control of the top-of-car operating device.

- 4.7.3, Anti-creep Leveling Devices, states that each elevator is to be provided with an anti-creep leveling device conforming to the following provisions:
  (a) It is to maintain the car within three inches of the landing irrespective of the position of the hoistway door.
  (b) For electrohydraulic elevators, it is required to operate the car only in the up direction.
  (c) For maintained pressure hydraulic elevators, it is required to operate the car in both directions.
  (d) Its operation is permitted to depend on the availability of the electric power supply provided that:
    (1) the power supply line disconnecting means required by 3.10.5 is kept in the closed position at all times except during maintenance, repairs, and inspections; and
    (2) the electrical protective devices required by 4.7.4(b) do not cause the power to be removed from the device.
- 4.7.4, Electrical Protective Devices, states that electrical protective devices, conforming to the requirements of 3.10.4, where they apply to hydraulic elevators, are to be provided and operate as follows:
  (a) The following devices are to prevent operation of the elevator by the normal operating devices and also the movement of the car in response to the anti-creep leveling device:
    (1) Stop switches in the pit
    (2) Stop switches on top of the car
    (3) Car side emergency exit door electric contacts, where such doors are provided
  (b) The following devices are to prevent the operation of the elevator by the normal operating device, but the anti-creep leveling device required by 4.7.3 is to remain operative:
    (1) Emergency stop switches in the car
    (2) Broken rope, tape, or chain switches on normal terminal stopping devices when such devices are located in the machine room or overhead space
    (3) Hoistway-door interlocks or hoistway-door electric contacts
    (4) Car door or gate electric contacts
    (5) Hinged car platform sill electric contacts
    (6) In-car stop switch, where permitted by 3.10.4(t)
- 4.7.5, Power Supply Line Disconnecting Means, states that they are to conform to the requirements of 3.10.5.
- 4.7.6, Devices for Making Hoistway-Door Interlocks or Electric Contacts, or Car Door or Gate Electric Contact Inoperative, states that they are to conform to the requirements of 3.10.7.

- 4.7.7, Control and Operating Circuit Requirements, states that they are to conform to the requirements of 3.10.9 and 3.10.12.
- 4.7.8, Emergency Operation and Signaling Devices, states that they are to conform to the requirements of Section 3.1.1.
- Section 4.8, Additional Requirements for Counterweighted Hydraulic Elevators, provides that they are to be roped so that the counterweight will not strike the overhead work when the car is resting on its fully compressed buffer. Counterweighted hydraulic elevators are to conform to the requirements of Section 3.2 where applicable. Where counterweights are provided, counterweight buffers are not to be provided.
- Section 4.9, Additional Requirements for Roped-Hydraulic Elevators:
- 4.9.1, Top Car Clearance, states that roped-hydraulic driving machines, whether of the vertical or horizontal type, are to be so constructed and so roped that the piston will be stopped before the car can be drawn into the overhead work. The top car clearance is to meet the requirements of 2.4.4.
- 4.9.2, Top Counterweight Clearance and Bottom Counterweight Runby, states that where a counterweight is provided, the top clearance and the bottom runby are to conform to the following:
  (a) Top Clearance is not to be less than the sum of the following:
    (1) The bottom car runby
    (2) The stroke of the car buffers used
    (3) Six inches
The minimum runby specified is not to be reduced by rope stretch.
- 4.9.3, Protection of Spaces Below Hoistway, states that where the hoistway does not extend to the lowest floor, the space below the pit is to be enclosed with permanent walls or partitions to prevent access.
- 4.9.4, Piston Stops, states that piston stops are to be provided to bring the piston to rest at either end of the piston travel from maximum speed in the up direction, under full pressure without damage to the driving machine, piston, piston joints, cylinder, cylinder couplings, or any other part of the hydraulic system.

    For rated speeds exceeding 100 feet per minute where a solid metal stop is provided, means other than the normal terminal stopping device are to be provided to retard the car to 100 feet per minute with a retardation not greater than gravity, before striking the stop.
- 4.9.5, Piston Connections, states that:
  (a) Equalizing Crosshead: Where more than one piston is used on the puller-type roped hydraulic elevators, an equalizing crosshead is to be provided for the attachment of the rods to the traveling sheave frame to ensure an equal distribution of the load to each rod.
  (b) Equalizing or Cup Washers are to be provided under piston rod nuts to ensure a true bearing.

(c) Piston rods of the puller-type hydraulic elevators are to have a factor of safety of not less than eight based on the cross-sectional area at the root of the thread of the material used. A true bearing is to be maintained under the nuts of both ends of the piston rod to prevent eccentric loadings on the rod.

- 4.9.6, Car Safety Devices, states that car safety devices conforming to the requirements of Section 3.5, except 3.5.2 are to be provided. Counterweight safeties are not to be provided.

- 4.9.7, Car Speed Governors, states that car speed governors conforming to the requirements of Section 3.6 are to be provided.

- 4.9.8, Sheaves, states that sheaves are to be cast iron or steel and are to have finished grooves for ropes.

    The traveling sheaves are to be guided by means of metal guides and guide shoes. The guide shoes are permitted to be equipped with nonmetallic inserts. Sheave frames, where used, are to be constructed of structural or forged steel and are to be designed and constructed with a factor of safety not less than eight for the material used. Single continuous straps (known as U-strap connection) are not to be used for frames or as connections between piston rods and traveling sheaves.

- 4.9.9, Slack-Rope Device: Roped-hydraulic elevators are to be provided with a slack-rope device and switch of the enclosed, manually reset type that will cause the electric power to be removed from the pump motor and the valves if the hoisting ropes become slack or are broken.

- 4.9.10, Suspension Ropes and Their Connections: All elevators, except freight elevators that do not carry passengers or freight handlers and have no means of operation in the car, are to conform to the following requirements:
  (a) Suspension ropes are to conform to the requirements of 3.12.1 through 3.12.3, 3.12.5, 3.12.8, and 3.12.9.
  (b) The minimum number of hoisting or counterweight ropes used for roped hydraulic elevators is not to be less than two.
  (c) The minimum diameter is to be 0.375 inches and the outer wires of the rope are to be not less than 0.024 inches in diameter. The term "diameter" where used in this section refers to the nominal diameter as given by the rope manufacturer.

The most common hydraulic elevator has a conventional configuration with a single below-grade cylinder directly below the car. Because of required excavation depth, height is generally restricted to four or five stories. A telescoping piston permits higher rises, at the cost of greater complexity. Combination roped-hydraulic systems allow the car to move farther than piston travel.

Hole-less hydraulic elevators, with two above-ground cylinders, are an option where high water table or bedrock preclude a conventional design. Where the site permits, the less complex conventional hydraulic elevator has been well-suited for low-rise, low-traffic installation. A downside is that they are less energy efficient than purely traction designs. High current draw when the pump starts under load places a greater demand on facility electrical resources, so for a new installation, alternatives should be weighed. The latest low-cost machine room-less traction elevators (see below) are strongly competitive in areas where previously hydraulic elevators were the clear choice.

Because of high startup current draw, in an outage emergency power may not be used to operate a hydraulic elevator, unless it is designed to do so. Typically, emergency power is used to lower the car to the next landing, and to open the doors. There the car rests until normal power is restored. In a low-rise building, occupants can use the stairs. In healthcare facilities, the emergency power system must be sized out to run the elevators throughout an outage.

Traditional elevator configurations include a machine room located at or below the lowest landing or above the top of the hoist. The machine room typically includes, for a traction elevator, separate electrical feeders for the motor (via VFD) and motion controller and for lighting, receptacles, and outlets in the machine room. For the motor, a dedicated disconnect must be located within sight in the machine room. Also in the machine room are the motion controller, VFD, motor with gearbox and related mechanism, and the drive sheave with pulleys and wire ropes. There may be a telephone, work table, and file cabinet for documentation. For a hydraulic elevator, the machine room consists of many of the same components. The difference is that rather than the type of motor and drive mechanism unique in a traction elevator machine room, the hydraulic elevator machine room houses an oil reservoir with submersible pump/motor and associated wiring and piping.

Machine Room-less Elevators

The machine room brings together many elevator components so they can be readily accessed for maintenance and servicing. The only downside is that valuable space within the building is not available for other essential services. To confront this problem, manufacturers developed the machine room-less (MRL) elevator, shown in Figure 2-4.

The MRL design was made possible by a new generation of smaller, lighter permanent magnet motors that permit installations consisting of the motor and associated components to be located in the hoistway without benefit of a machine room.

MRL hoisting methods allow a reduced sheave-to-rope ratio of 16:1 as opposed to the 40:1 ratio in the conventional traction elevator configuration. At the smaller ratio, a more flexible, higher-strength wire rope is used.

The MRL design incorporates motor, drive sheave, counterweight, and wire ropes as in both the geared and gearless traction elevators. In MRL elevators, the gearless drive is preferred although either is possible. The MRL components are located in a space above the hoistway except for the motion controller, which may be located

FIGURE 2-4   The machine room-less motor is located on top of the car or elsewhere in the hoistway. (*Wikipedia*)

in a locked cabinet in the top floor hallway adjacent to the shaft door. MRL elevators may be either traction or hydraulic. MRL elevators do not have a fixed machine room at the top of the hoistway. Instead, the traction hoisting machine is installed either on the top side wall of the hoistway or on the bottom of the hoistway. The permanent magnet motor works in conjunction with a VFD. This design eliminates the need for a machine room and saves space. While the hoisting motor is installed on the hoistway side wall, the main controller is installed on the top floor next to the landing doors. Most elevators have their controller installed on the top floor, but some are installed on the bottom floor. Some elevators have the hoisting motor located at the bottom of the elevator shaft pit. This is called a bottom drive MRL elevator. The controller cabinet may be installed in the door frame. MRL elevators sometimes use flat steel belts instead of wire ropes, permitting a smaller hoisting sheave. Machine room-less elevators in mid-rise buildings usually serve less than 20 floors. The traction mechanism may be located under the elevator cab as in some Schindler designs. Like the traction version, machine room-less hydraulic elevators do not have a fixed room to house the hydraulic machinery. In the MRL design, hydraulic machinery is located in the elevator pit. The controller is located on a wall near the elevator on the bottom floor. MRL hydraulic elevators like the traction models require less space.

Rather than in a machine room, most components in MRL designs are in the shaft. The motor and drive mechanism may be on top of the car, under the car, at the top of the shaft, or at the bottom of the shaft. The motion controller is frequently located in a locked cabinet in the top-floor hallway adjacent to the hoistway door. Except for their compact size and unusual locations, components are similar to those in conventional traction or hole-less hydraulic elevators. Kone introduced the MRL design in 1996 and it is currently offered by many manufacturers.

In addition to freeing up valuable space, the MRL design uses less energy and initial cost is significantly lower. A significant disadvantage is that maintenance and servicing are more difficult, and workers have been injured. (Imagine doing dynamic vibration testing on the motor on top of a moving car!)

Double-Deck Elevators

In a double-deck elevator, there are two attached cars, one on top of the other. They move together in the same shaft. The great advantage in a many-story building is that two adjacent floors can be served simultaneously, with half as many stops. The capacity of each shaft is doubled, cutting down on dedicated floor space on each story.

In some designs, one of the cabs serves as a freight elevator. During peak traffic periods, it becomes a second passenger car.

Worldwide, many double-deck elevators have been built, as many in Asia as in Europe and North America combined.

Special-Purpose Elevators

In densely populated areas where space is limited, multi-level parking lots have inclined ramps so that users can drive to the desired level. Some of these facilities have automotive elevators that carry cars and passengers to their destinations. Most of these are hydraulic elevators, and they must be rated for large loads to safely accommodate a heavy vehicle loaded with passengers and luggage.

Closely related are aeronautic elevators on aircraft carriers. Here considerable space is needed on deck for the runway plus nautical equipment, and that leaves room for only a few aircraft. Most of the 100 or so jets are stored in below-deck hangars, which is convenient for servicing and general maintenance. But how are they moved back and forth? Early aircraft carriers experimented with various methods such as moving the aircraft up and down inclined ramps, and using cranes to lift them. In the end, a nineteenth-century solution was adapted for this twentieth-century need, the hydraulic elevator.

In a typical aircraft carrier, most of the jets that are not in use are stored in the hangar bay, which is located two decks below the flight deck, directly under the galley deck. The hangar bay may be 110 feet wide, 25 feet high, and close to 700 feet long, over two thirds the length of the aircraft carrier. Besides aircraft, there is a large maintenance area with spare engines, extensive tools, and areas open to the outside for testing engines. Various sections are separated by sliding fire doors. There may be four gigantic aluminum hydraulic elevators, each capable of lifting two 30-ton jets simultaneously.

Stage elevators are essential for large theatrical productions. They permit elaborate settings to be assembled in advance, ready for action, and raised into place between scenes. Similarly, large orchestras complete with musicians and heavy pianos can be instantly deployed. These elevators are invariably hydraulic, with heavy steel-framed floors supported by massive pistons. Radio City Music Hall has four large stage lifts behind which are three smaller ones, so that sections of the set and orchestra can be moved as needed.

Residential elevators are used to transport elderly and handicapped persons between floors in their own homes, often permitting partially disabled persons to live at home for many years without relocating to a ranch-style building. These home lifts are permitted to be less complex, and they are generally slower and less powerful than commercial elevators, so the cost is far less. Safety systems such as shaft and car door interlocks, safeties, and emergency phones are required. ASME A17.1, Section 5.3 covers residential elevators. This category does not include elevators in multi-family occupancies.

Dumbwaiters are intended to carry light freight between floors, never human passengers. They are frequently installed in restaurants and hotels, to connect kitchens to dining rooms on higher levels. Because of the lighter, inanimate loads, some of the usual safety features are not required. The platforms are much smaller than those in passenger and conventional freight cars, with three-foot door heights. At each landing there is a control panel, permitting calling, door control, and choice of destination.

The Paternoster consists of a series of moving compartments. It resembles the humanlift, found in multi-story industrial buildings, where the rider stands on a small platform, gripping a handhold.

The scissors lift is a mobile work platform that allows maintenance workers and painters to easily access outside walls and inside walls in industrial building that have high ceilings. Scissors lifts are used extensively by electricians in servicing light fixtures in high-bay areas. They can also be used as long-term fixed elevators where there is no space for pit, machine room, counterweight, or hydraulic cylinder.

Rack and pinion elevators are powered by a motor-driven pinion gear. They are often installed on the outside of a building under construction, and can easily be moved to the next work site.

Belt elevators are used extensively to move loose material such as grain and coal. Typically the loads are moved up inclined planes. Pit mines make extensive use of these elevators. Vertical elevators in underground mines, used to transport materials, equipment, and workers, are often large, powerful, and challenging in terms of design, installation, and maintenance. Electrical systems are sometimes in areas classified as hazardous.

Elevator Manufacturers

A survey of elevator types should include a description of the major elevator manufacturers. There are relatively few compared to the number of household appliance or automobile manufacturers. This is because elevator systems are very complex. They vary widely in height, with shafts, traveling cables, and steel ropes sometimes thousands of feet long. Often there are group installations of multiple powerful high-voltage electric motors and elaborate control and communication systems. The entire assembly has to conform to detailed safety codes, which vary depending on the location and elevator type.

Here is Wikipedia's list of current elevator manufacturers. We will focus on a few of them:

- Acorn Stairlifts
- Aichi
- Anlev Elex
- Anton Freissler
- Canton Elevator Incorporated
- Cibes Lift
- Delaware Elevator Manufacturing
- Delta Elevator
- FUJIHD
- Fujitec
- GEDA USA
- Hitachi
- Hosting Elevator
- Hyundai Elevator
- KLEEMANN
- Kone
- LG Elevator
- Liftech SA
- Marshall Elevator
- MEI-Total Elevator Solutions
- Mitsubishi Electric
- Orona Group
- Otis Elevator Company
- Schindler Group
- Servas ascensores
- Sicher elevator
- Sigma Elevator
- Symax Lift
- Stannah Lifts
- ThyssenKrupp
- Toshiba
- Ulift

Hitachi is moving forward with its UAG Series SN1 and OUG Series ON1 machine room-less elevator systems. Hitachi offers advanced functions, some standard and some optional, including simplex, duplex, and group control, automatic return function, independent and attendant parking and rush-hour schedule operation, interphone system, floor lockout and door nudging, abnormal speed protection, and out of door-open zone alarm.

Additionally, when the car stops out of the door-open zone, it will move at slow speed to the nearest floor to release passengers. In the event of door overload, such as when passengers get their fingers, hands, or personal belongings caught in the door, the system automatically senses this and either recloses or reopens the doors to prevent injury.

Micro-leveling automatically corrects the car level when there is a difference between the car and the landing floor. In the event of a power failure, an emergency light inside the car is automatically activated. A battery-powered emergency supply allows the operation of light and alarm bell. A multi-beam door sensor installed at the edge of doors, in the event that beam paths are obstructed, will keep the doors open.

As of 2018, Hyundai has produced over 28,000 elevators. The company states that safety, ride quality, and space efficiency are combined in optimal elevator systems that will enhance the value of buildings. Speeds are up to 18 meters per second and maximum passenger capacity is 30. Hyundai manufactures observation, hospital, and freight/automobile elevators with a maximum rated load of 5,000 kg. All products can be custom-made to suit each building.

Kone, a Finnish elevator manufacturer founded in 1910, conceived the machine room-less elevator in 1996. Over the years throughout the world Kone has bought and sold many corporations. At various times it has invested heavily in forest machines, cargo handling, and tractors. Currently, it has narrowed its focus and expanded its elevator business. Due to environmental and energy efficiency concerns, Kone announced in 2007 that it would no longer manufacture hydraulic elevators.

In 2011, Kone built a new headquarters building named the Kone Center in Moline, Illinois.

In 2013, the firm introduced Kone UltraRope, designed to replace steel rope in traction elevator systems. It will permit building heights up to one kilometer. UltraRope is lightweight, with a high-friction coating and consequently reduces energy consumption in high-rise applications. With a higher resonant frequency than steel rope, cable sway is reduced in long runs, with less possibility of cable and shaft damage.

Mitsubishi Electric also offers machine room-less elevator systems. Elevator Group Control System ΣAI-2200C incorporates fuzzy logic to provide comfortable elevator operation and ride under changing usage conditions. The system incorporates intuitive control to provide smooth operation. When a hall call button is pressed, the optimum car responds based on waiting time, travel time, current car occupancy, and projected energy consumption.

Otis is the world's largest elevator manufacturer. We told the story in Chapter One of Elisha Otis's invention in 1852 of safeties, which lock the car to guides fastened to the hoistway walls so that it can't fall into the pit if the steel ropes fail.

The Otis brand is well-known throughout the world. The firm maintains extensive research and testing facilities, and watches its competitors, who in turn watch Otis for industry-changing innovations, such as Kone's machine room-less systems.

Schindler Group, like many of its competitors, has enthusiastically embraced the MRL concept. The company currently employs over 58,000 people in the United States, Switzerland, India, Spain, Slovakia, Brazil, and China. Schindler was founded in 1874 in Switzerland and soon began manufacturing many types of machines including elevators. It rapidly expanded throughout Europe, the United States, South America, and China.

Schindler has introduced Miconic 10, a proprietary control system. Passengers enter their destinations in a wall-mounted keypad while waiting for a car. The system groups riders with the same destination to a specific car, vastly increasing efficiency and reducing waiting time.

ThyssenKrupp has continuously embraced new technologies such as double-decker cars, rope-less horizontally-moving cars, moving walks, heavy-machinery freight, and vehicle elevators. The company has 670 subsidiaries.

STUDy QUESTIONS

1. The most common type of elevator is:
   A. machine room-less
   B. hydraulic
   C. traction
   D. aeronautic

2. Traction elevators are characterized by:
   A. water power
   B. freight only
   C. an operator
   D. steel ropes

3. The gearless design permits car speeds in excess of:
   A. 100 fpm
   B. 500 fpm
   C. 1000 fpm
   D. 2000 fpm

4. Compensation is needed when vertical travel is greater than:
   A. 100 feet
   B. 250 feet
   C. 500 feet
   D. 1000 feet

5. Compensation consists of:
   A. a greater amount of insurance
   B. hanging chain to add weight
   C. good judgment on the part of technicians
   D. higher speed for a smoother ride

6. Steel belts can be used rather than steel rope. They require:
   A. frequent lubrication
   B. more power to run
   C. less material weight
   D. a skilled operator

7. Traction elevator motors with gearless drives:
   A. are smaller
   B. operate at lower RPM
   C. require more maintenance
   D. can be noisy

8. Roping configurations in a traction elevator:
    A. consist of suspension ropes attached to the top of the car
    B. may be underslung
    C. pass down to the counterweight
    D. any of these

9. The gearless design requires:
    A. higher car speed
    B. a smaller sheave
    C. frequent lubrication
    D. frequent rebuilding

10. In a regenerative drive system, energy is returned to the grid:
    A. when a full car is ascending
    B. when an empty car is descending
    C. when a full car is descending
    D. when the car is stopped

For answers, go to Appendix A.

AC AND DC ELECTRIC MOTOR MAINTENANCE, VFD TROUBLESHOOTING, AND DIAGNOSTIC PROCEDURES

Almost all modern elevators, traction and hydraulic, are driven by electric motors. Until the development of the VFD in the 1960s, electric elevators invariably made use of the DC as opposed to the AC motor because the former permitted simple, smooth speed control. That was essential for stopping the car at individual floors and also for running in inspection mode.

Since the introduction and widespread use of AC motors in conjunction with VFDs, three-phase induction motors have been used in most new construction and major rebuilds, although in the latter, a case can be made for retaining a good working DC motor. Parts and entire replacement motors are readily available and the most common repair, brush replacement, is not difficult.

In this chapter, we will discuss DC and AC motor maintenance, as shown in Figure 3-1, beginning with some comments that apply to both.

Good Motor Maintenance

Some first signs of motor trouble are failure to start, excessive or unusual noise, reduced power, odor from insulation breakdown, overheating, and tripping out. These symptoms often occur in combination, although at first only one of them may be apparent.

The goal of a good maintenance program should be early detection rather than running to failure. In an industrial facility, large hotel, or office building, maintenance persons typically tour the building at least once per shift, checking critical areas including sprinkler systems, central fire alarms, elevator machine rooms, and the like. At important locations, it is beneficial to have a maintenance log, usually a clipboard

FIGURE 3-1   AC elevator motor.

mounted on a nearby wall. There should be space for date, time, metrics such as temperature or air pressure, comments, and the worker's initials. That worker should also verify that important indicators are in line with previous entries, that ventilation fans and all lights are operating, and the area is not being used as storage or a catch-all that would impede air circulation and ready access. Strategically placed thermometers are helpful in detecting heat rise. Thermal imagers, shown in Figure 3-2, available from Fluke, Grainger and Amazon, are excellent tools.

Machine Room

The elevator machine room, in a conventional installation, provides access to the motor. It also contains the motion controller, VFD, associated wiring, and electrical disconnect within sight of the motor. It's very helpful also to have a work table and file cabinet to store documentation, and a telephone with elevator and motor manufacturers' tech help numbers posted conspicuously.

FIGURE 3-2   Thermal imager. Nothing could be easier.
Just look into the viewer for hot spots. (*Fluke*)

Included in the documentation should be a service history that lists any atypical maintenance log entries and corrective measure taken. Brush and bearing changes, commutator inspection and servicing or replacement, and motor teardown, lubrication, and cleaning should be noted, as well as Megger test results.

Before doing motor work, the elevator must be put out of service. This means communicating with building managers and placing appropriate signs or barriers at all floors. If possible, the car should be parked adjacent to the machine room. Except for external cleaning and thermal imaging, the disconnect should remain in the off position, verified by means of a known good voltmeter. Then, the disconnect can be locked in the off position.

It is normal in a DC motor for the brushes, shown in Figure 3-3, to wear. They ride on the commutator and there is a certain amount of friction. Additionally, electrical current is conveyed through the brushes to the commutator, and this also creates wear. The unavoidable combined effect is that in time the brushes become shorter. To compensate, they slide in their holders. Springs keep them pressed lightly against the spinning commutator.

In smaller motors, these springs also conduct current to the commutator. Brushes for larger motors have flying leads. Where the brushes contact the commutator, there is a certain amount of sparking, which with increased wear becomes brighter. Eventually this sparking comes to resemble a bright continuous flame. By that time it has gone too far and it is certain the commutator is being damaged.

The motor documentation should state the minimum brush length or wear limit, and it is important to not run the motor beyond this point. If this happens or if the brushes are not pressed lightly against the commutator due to springs that have lost

FIGURE 3-3 Brush replacement is a routine part of DC motor maintenance. (*Amazon*)

their tension, an arc will develop. The longer the arc, i.e., larger the gap, the more heat is created, and soon the commutator will be damaged. If this damage has not become excessive, the commutator can be renewed by turning it on a lathe and then using a special tool that resembles a hacksaw blade with a handle to cut back the edges of each insulating layer between commutator segments. This work, of course, involves a motor teardown. It is to be emphasized that a set of brushes is far less expensive than performing commutator work.

You might be tempted, if the correct brushes are not on hand, to cut down larger brushes or fabricate holders to make them work. The problem here is that in addition to physical dimensions, brushes have other characteristics. Often conductive metal particles are added, which influence their electrical behavior. The best thing to do is to obtain spare brushes in advance so that you have them when needed.

If new brushes were installed, the DC motor may be back to normal. Before restoring power, it is a good idea to do a preliminary check of the bearings. A total evaluation would require motor disassembly, but for now the object is to check for side play. Simply attempt to move the shaft from side to side. Any discernable side play relative to the motor housing indicates that bearing failure is imminent. It is preferable, of course, to replace front and rear bearings rather than running the motor to failure.

Bearings

It is not unusual to replace bearings one or more times in the life of the motor. It is said that bearings under optimum conditions will last 100,000 hours, but there are numerous factors that can affect this figure. Premature failure may be due to over-loading the motor or running it above rated speed, excessive heat (ambient or from within), inadequate or over-lubrication, prolonged motor vibration, or electrical current flowing through the bearing.

Damage that begins in one location in the motor can quickly spread. At times windings may be blackened and the primary cause may be defective power quality, environmental moisture or dust, or winding insulation failure. But the trouble often starts with faulty bearings.

Mechanical distortion including dents and scratches or out-of-round balls or races can cause bearing and overall motor heating. If the shaft is worn, a motor shop can perform flame-spray metalizing to apply new metal to the shaft, which is subsequently turned down on a lathe to original specifications.

Motor Shaft and Bearing Current

As discussed earlier, motor shaft and bearing current damage is a major problem in large electrical motors, especially AC but also DC. The best plan is to detect this current at the outset, when the motor is first put in service, and eliminate it before bearing and complete motor damage results.

Motor shaft voltage and bearing current damage is more prevalent in AC induction motors operated in conjunction with VFDs than in DC motors. The reason is that the pulse-width modulated current with its fast rise and fall times equates to high-frequency electrical energy that couples between windings and shaft. DC electrical energy less frequently shows up as shaft current, but when it does the motor bearings can be likewise degraded.

VFDs are the source of common-mode current that is conveyed through the power conductors and infiltrates the motor. Without the presence of this damaging current, motor bearings, if properly lubricated, require minimal servicing. If there is bearing current, they may be destroyed in a few days.

High-frequency spikes, inevitably present in pulse-width modulated power from a VFD, capacitively charge the cable and conductive elements in the motor that are insulated from one another. Where there is bearing current the swirling grease in the bearings is partially conductive and therefore subject to resistive heating as well as arcing at the stationary and rotating metal surfaces.

Bearing current can be detected by instruments such as the Fluke Motor Drive Analyzer shown in Figure 3-4, using the Motor Shaft Voltage function.

Shaft voltage should be measured after the motor is energized, rotating, and warmed up to normal operating temperature. Measuring voltage on the rapidly rotating shaft is facilitated by a shaft voltage probe, which has an attached small conductive brush that contacts the spinning shaft. The ground reference lead is clipped to the motor enclosure.

Bearing current damage can be seen in the appearance of the grease, which rapidly becomes dark due to burnt metallic particles. The bearing surfaces, including balls, prematurely acquire a darkened and/or frosted appearance, or a geometrical stripping similar to washboarding in a heavily-traveled gravel road.

When symptoms appear, the bearings should be replaced. Shaft voltage and bearing current can be eliminated by creating a jumper around the bearing. This jumper should be between the motor enclosure, which is at ground potential, and the shaft. The problem with this method is that the shaft is rotating, so contact has

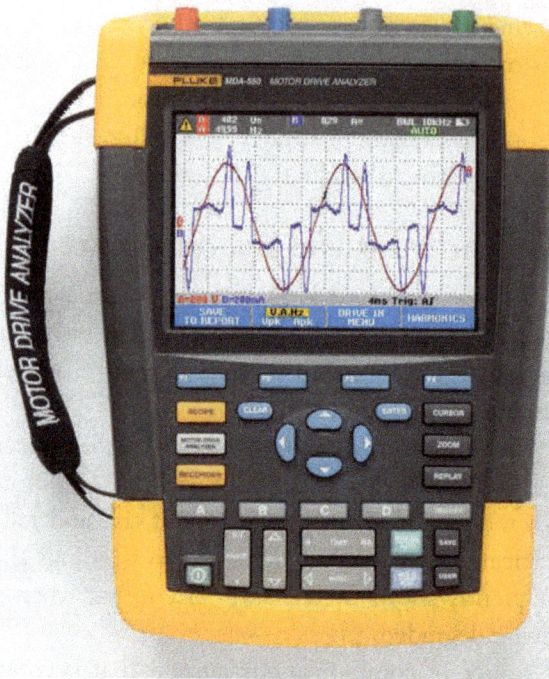

FIGURE 3-4   Fluke MDA-550 Motor Drive Analyzer, 4 channel, 500 MHz. (*Fluke*)

to be made by means of a conductive brush, which is subject to wear and could fail. The other method is by installing CoolBlue common-mode chokes, also known as inductive absorbers. They last indefinitely and are easily installed by disconnecting power cables, slipping the chokes over them, and reattaching cables. (For details, see www.Coolblue-mhw.com.)

In elevator operation, run to failure, as opposed to preventive and predictive maintenance, is not a viable option. In a hotel or high-rise apartment or office building, it is essential to detect potential mechanical and electrical problems before they occur, thereby avoiding an outage. Nobody is perfect, but an extended elevator outage should be a once-in-a-century event, not a yearly occurrence. Reliable motor operation is essential. The following sections cover things you can do to ensure trouble-free operations.

Vibration Analysis

Vibration analysis consists of taking baseline readings when a new motor is placed in service and comparing them to subsequent periodic readings. In that way, damaging trends can be detected and corrective measures taken before failure occurs or expensive repairs and downtime result. The rationale for vibration analysis is that it is an early indicator of the health of rotating machinery, especially large electrical motors. Equipment condition can be monitored without motor teardown and internal anomalies can be identified and measured.

As in an oscilloscope, a display with time represented on an X-axis and amplitude on a Y-axis is created within instrumentation that is linked to the motor. A unique vibration signature can then be compared to previous readings to reveal what is happening inside the enclosure. Excessive vibration is indicative of imbalance, misalignment, looseness, and incipient bearing failure.

Vibration data is gathered by either or both of two methods. It can be measured at the rotating shaft relative to the enclosure or at the enclosure relative to an absolute reference, such as the building's structural steel. Two proximity probes are used, one at each location.

A large proportion of motor failures can be traced back to bearing problems, long before they fail catastrophically. And most bearing problems are due to improper lubrication. Insufficient lubrication is a direct cause of bearing damage because the dry bearings heat up and the bearing surfaces lose metal and may eventually seize.

Over-lubrication also causes damage. Excessive grease within the bearing heats, tries to expand, and, since it is confined, exerts pressure impeding rotation and creating more heat. Also, excess grease can find its way into windings and cause insulation to deteriorate.

Manufacturers' documentation provides guidance regarding frequency and amount of lubrication for optimum performance. A graduated or calibrated grease gun ensures the maintenance technician knows precisely how much grease is added.

The air gap between rotor and stator can be checked using a feeler gauge. Top, bottom, and side measurements should be recorded. If there is bearing wear, the top gap will increase and the bottom gap will decrease. Both side gaps usually increase.

If budget permits, having spare motors on hand is a great help in avoiding an extended outage. In a group installation, if all motors are the same, a single spare should suffice. The shaft of a spare motor on the shelf should be rotated every 90 days. The purpose is to avoid outer race deformation due to the weight of rotor-shaft assembly plus the commutator in a DC motor. Periodic vibration readings should be taken under similar conditions (speed, loading, location of proximity probes) in order to maintain a consistent database.

Bearing replacement is a critical and exacting operation, more so in larger motors. Machinists' levels and gauges must be used to achieve proper alignment relative to X, Y, and Z axes. Shaft fit is critical. Always check shafts to maintain specified tolerances. Lock the bearing to the shaft correctly. Both bearing inner races are preheated to a uniform 110° C to shrink-fit them to the shaft journal. Bearing shoulders and journals need to be cleaned and polished using emery cloth, with all particles removed prior to heating.

In motor operation, stator faults can lead to total motor failure. The integrity of insulation between individual turns is critical, and if it is compromised, the motor will overheat. In a larger motor, there are significant forces that are exerted on the stator coils, especially when a fully-loaded motor starts. High magnetic force on the stator coils causes them to vibrate. The mounts are stressed and tend to loosen. Additionally, worn bearings and misalignment can cause the rotor to contact the stator coils, resulting in grounded coils and damage to rotor and stator, not to mention severe bearing current.

An array of motor problems, including poor air flow, unequal voltage and current in the three phases, worn bearings, degraded insulation, and shaft misalignment all show up as heat. Early warning of these and other impending problems is provided by the thermal imager. Maintenance workers should routinely use this instrument to examine the elevator motor and, while they are at it, to inspect the motion controller and high-current terminations at the motor disconnect and at the motor. The ideal procedure is to image these components at initial installation and periodically thereafter.

The good news is that infrared imaging is quick and easy, and it is far more sensitive for many types of fault detection than vibration and motor current signature analysis, both of which, however, are needed for comprehensive analysis.

Thermal imaging works in detecting internal motor faults without opening the motor housing. Internal partial short circuits to ground and between windings are readily apparent by imaging from outside of the enclosure. Also, it is revealing to compare identical motors loaded the same, as in an elevator group installation. Rear and front bearings when not side loaded should dissipate equal amounts of heat.

In an induction motor, stator winding faults and insulation deterioration are revealed as temperature rise. Since the rotor functions as a transformer secondary, any faults there produce localized heat rise.

Motor Drive Alignment

When an electric motor is coupled to driven machinery or to an intermediate gearbox, shaft alignment is critical. The two principle types of traction elevators, geared and gearless, have electric motors that are coupled to the driven machinery in different ways. In geared traction elevators, the motor drives a gearbox, which is a single-speed transmission, and the gearbox drives the wheel, moving the ropes. Gearless traction elevators are similar, but there is no gearbox. The motor drives the sheave. In both types, drive-train alignment is very important. Shaft alignment can be defined as the degree to which two or more machines are positioned such that the axes of rotation of all shafts are collinear when running at rated speed and normal loading. There are exceptions. Some gear couplings require very slight misalignment to maintain correct lubrication.

There are many causes of poor alignment. Some of them are the result of defective initial installation, while others develop days, months, or years later. Typically, motors and driven machinery are bolted to a concrete footing or floor. If this support structure cracks or moves, the drive will be thrown out of alignment. Other causes are loose, stretched, or broken foundation bolts or loose nuts, loose shims or dowels, loose or broken coupling bolts, unequal or excessive vibration, improper or defective couplings, and bent or cracked shafts.

Misalignment causes accelerated drive train component wear, decreased speed and less than optimum performance, increased maintenance costs including labor and parts, increased energy costs, compromised elevator safety, and possible down-time and outage. For these and other reasons, proactive maintenance is highly advisable. Periodic thermal imaging with written records and review to uncover damaging trends is a low-cost, high-benefit practice. Additionally, bearing noise usually signifies that bearing failure may occur in the near future. Before or after these symptoms appear, alignment can be checked using a straightedge and feeler gauges in conjunction with a machinist's level. Greater precision may be attained using laser optics instrumentation.

Correcting misalignment consists of determining the amount of error, then moving the motor, gearbox, or driven machinery as required and anchoring it down securely. Since we are usually dealing with small corrections, they can be accomplished by lifting the machine(s) a slight amount, then inserting metal shims.

The machine can be moved horizontally by means of jack bolts or hydraulic tools, as opposed to hammers, which are not suitable for exact positioning and may damage bearings or even crack castings. (Alignment is a gentle art.) Once correct alignment is achieved, it is necessary to anchor everything down so that in the fullness of time despite vibration and continuous use, correct alignment is maintained. In a new installation, the driven machinery should be installed first, then mounting bolts torqued to specifications found in Machinery's Handbook. Then, the motor is positioned and fastened in place. If there is a gearbox, it should be aligned to the driven machine and finally the motor aligned to the gearbox.

Next on the agenda is to check for accurate positioning of the machine couplings.

Even if motor, driven machinery, and gearbox are aligned correctly vertically and horizontally, coupling halves must also be true to ensure correct balance. Once more, check machinery base, mounting surfaces, feet, and pedestals. File burrs and metal irregularities from all mounting surfaces and use no more than three metal shims at each mount.

Electrical Issues

If a motor fails to start or its over-current devices trip out, and assuming alignment, ventilation, and lubrication levels have been considered, it may be appropriate to begin extensive electrical testing. Some of this is done with the motor and driven equipment running. For other tests, the motor is powered down, motor conductors disconnected at the motor terminals, and driven load disengaged. In some instances the motor may appear to malfunction, but the actual problem is binding of the load.

The most basic tests involve power quality. Usually the power source is supplied by the utility to the building's electrical service. In this discussion, we will begin at this upstream location and proceed downstream to the motor. This is the logical way to talk about the system, but experienced technicians generally begin at a halfway point so that large portions of the system can be eliminated as faulty at the outset.

We discussed electrical safety in the Introduction, but now we'll take a closer look.

To take high-level voltage and current measurements, it is generally necessary to approach and open an electrical enclosure or terminal box. This has to be done correctly in order to avoid hazards from electrical shock and arc flash.

Elevators are typically powered by three-phase AC induction motors in conjunction with VFDs. A common line voltage is 480 volts. To an electrician this is very high voltage while to a utility line worker it is low voltage. In fact, the boundary between high and low voltage is not universally defined. But 480 volts can arc over an inch through dry air. The human body can then become a conductive path to ground with lethal consequences. To avoid this hazard:

- Be aware that all objects including electrical workers have an inherent ground potential that varies with their mass. A bird has a very small mass and can sit with impunity on an uninsulated high-tension wire. Your inherent ground potential is greater, and it drastically increases if you are in contact with a large conductive object that can serve as ground, and more so if you are in contact with an actual earth ground. If you are immersed in water and touch a live wire, a very low-impedance circuit is formed with grave consequences. Accordingly, electric shock hazard can be partially mitigated by limiting your contact with ground. Electronic technicians often place a thick dry rubber mat on the floor at their workbench, and they use a wood or plastic, not metal, stool. This can lessen the severity of a shock but not eliminate it altogether.
- When measuring voltage higher than 240, it is advisable to wear high-voltage lineman's gloves, available at Amazon.com for under $100. Utility workers periodically inflate them with compressed air to detect any pinhole break. The slightest puncture can allow electrons to stream in, seeking ground. These gloves should be stored in a protected location and should never be used as work gloves.
- When taking voltage measurements, be sure that you are not overloading the instrument. Voltage limits are invariably printed prominently on the enclosure of a multimeter, oscilloscope, spectrum analyzer, or other measuring instrument. High-voltage probes have large barriers that prevent arcing and slipping of the hand.

These are some protective measures, but of course there are many ways in which a worker or passerby can be exposed to electric shock, so it is essential that you receive training in avoiding these hazards, and that you secure any required licensing. A good policy is to take multiple protective measures. For example, even if you know the power is off, do not touch potentially live conductors.

Another entirely different hazard is arc-blast. A worker can be injured even when an electrically energized conductor is not contacted and current does not flow through the body. Under some conditions, electrical injury other than shock can be just as severe. Either of these hazards can be fatal.

The classic case of arc-fault is when a conductive object such as a wrench is dropped so that it simultaneously contacts a live terminal and a grounded surface. There will be an electrical arc capable of injuring or killing nearby persons. Its severity is proportional to the amount of available current. The amount of current conforms to Ohm's Law:

$E = IR$ or, transposed, $I = E/R$
Where E = electromotive force in volts
    I = current (originally called intensity) in amps, and
    R = resistance in ohms

The problem in an electrical fault circuit is that R takes on a very low value, approaching but never reaching 0.

As you can see from the above equation, as R approaches 0, I approaches infinity. R never reaches 0 because it always equals the sum of the resistances or impedances in the wrench or arcing object, branch circuit conductors, bus bars and over-current devices, service drop conductors, utility lines, transformers, and generators.

In a commercial or industrial facility, the impedance is purposely kept low so that large motors and the like won't experience voltage drop, causing them to over-heat. The way impedance is kept low is by installing large transformers, conductors, and over-current devices, but this is at the cost of having high available fault current.

Utility engineers inform the facility electrical department of the available fault current at each service location, and these amounts are posted at the panels. Depending upon how close the worker approaches this type of switch gear, various protective clothing and devices must be used, as mandated by the Occupational Safety and Health Administration (OSHA).

Working on switch gear and terminations with high available fault current requires specialized training and techniques. Because elevators sometimes incorporate large motors (many of them in group installations), it is important to be aware of the extraordinary potential hazards in high-voltage and high available fault-current installations.

A fundamental principle in electrical work is to be sure it is de-energized where possible before performing maintenance or repair operations. Obviously this is not always possible. For example, voltage and current measurements as well as vibration analysis require that the equipment be energized.

When de-energizing a system, it is prudent to lock out the disconnect. Each electrician has a specially designed lock and key that will prevent the power from being turned on prematurely.

A later chapter focuses on power quality. For now, we will describe some preliminary measurements. Since most elevators have three-phase motors operated in conjunction with VFDs, we'll begin by looking at the anatomy of a three-phase system.

Diagnosing Three-Phase Circuits and Motors

Almost all motors five horsepower and over are three-phase. Most utility lines are three-phase. It is a simple matter to power a single-phase transformer and building from a three-phase utility line. One of the three phases plus ground feeds the transformer and building. That is the way most residences are served. Some people think this is two-phase power since there are two opposing phase wires. However, the two lines are in phase even though the instantaneous polarities are opposite one another. True two-phase power is rarely used. Most residential power is single phase.

Medium and large commercial and industrial users require three-phase power. All three utility lines are fed to the building through a step-down transformer. Driving through a commercial area that has three-phase aerial services, you will see three heavy lines, plus a fourth grounded neutral, going to a weather head at each building where they disappear from view going into conduit. Sometimes you will see six insulated service conductors. Here the conductors are paralleled in pairs, permitting smaller wires. The three phase conductors pass through a meter and main disconnect, thence to the entrance panel.

Rather than a double-pole main breaker and two bus bars as in a single-phase entrance panel, the three-phase box has a triple-pole main breaker and three bus bars. Each phase wire goes to the main breaker and through it connects to one of the bus bars. Neutral and grounding are as in a single-phase entrance panel.

Three-phase branch circuits receive three-phase power by means of three-pole breakers, which bolt onto all three bus bars. The three branch-circuit conductors are run along with a green or bare equipment grounding conductor and neutral where required to motors, other loads, and load centers throughout the building. Motors and large three-phase loads are supplied by dedicated circuits.

The convenient thing about three-phase power is that single-phase circuits can be derived from these same entrance panels and load centers, with no other equipment. You just bolt on single- or double-pole breakers (depending upon the voltage) and wire to the loads.

Wiring three-phase motors is easier than wiring single-phase motors because for a given horsepower the conductors are smaller. Bring the three wires plus equipment ground through the controller to the motor. Reversing any two of the wires will reverse the rotation, although this will not happen at the input of a VFD since the three-phase power is rectified internally. To determine rotation, use a three-phase rotation meter. Alternatively, for something like a fan, many electricians do trial and error. But beware—some pump seals are destroyed when the direction of rotation is wrong.

When a three-phase induction motor that powers an elevator is running hot, tripping out, or exhibits other faults, the first thing to do is check power quality at the VFD input. This measurement can be made at the input terminals or at the main disconnect. Actually, two measurements should be taken, one with the elevator running (if possible) and one with it not running. The supply voltage should not drop significantly when the motor starts up.

At the VFD input, the three phase-to-phase voltages should not vary more than one percent with no load. If greater variation is measured, work back toward the entrance panel. Has someone bugged off a single-phase load from what should be a dedicated circuit? Another possibility is a weak termination somewhere along the line, or damaged cable. For this kind of work, a thermal imager with load connected is useful. Also, especially if the conductors are aluminum, it may be a good time to retorque all terminations and verify that corrosion inhibitor was used at the initial installation.

To gain further insight, while enclosures are open, check current readings using an electrician's clamp-on ammeter. For these readings, of course, the motor must be running and preferably fully loaded, i.e., moving the car. If, suddenly, an imbalance appears, you may have a problem anywhere along the line to the motor, likely the windings or a termination.

If the readings are good, go to the VFD output and take similar readings. The problem here, however, is that a standard true RMS multimeter will not provide a reliable reading. The reason is that the VFD output to the motor is not a conventional sine wave as provided by the utility. It is a PWM waveform synthesized internally in the VFD inverter section by the IGBTs.

To realistically measure this three-phase VFD output voltage, a good instrument is the Fluke 87V multimeter, shown in Figure 3-5, which has an internal selectable low-pass filter.

FIGURE 3-5   The Fluke 87V multimeter is optimized to measure VFD output voltages. (*Fluke*)

This multimeter permits the user to take reliable voltage, current, and frequency measurements at the drive output or motor input terminals. If there is a difference in the readings, most likely the cable between VFD and motor has become damaged, or the problem is in the terminations at either end. Take a look with the thermal imager.

In these measurements, again we are looking for phase imbalance. If there is greater than one percent difference phase-to-phase, and a cable defect has been ruled out, we have to look to the motor or the VFD.

The VFD is easy to troubleshoot. As a preliminary, do a visual survey and take a look with the thermal imager. Under load, the sixteen components (typically six IGBT's, two capacitors, two inductors, six diodes) should all be uniformly warm, none hot and none cold. A cold component is open. A hot component is shorted. A warm component may be normal. Visually, none of the components should be burnt, leaking fluid, cracked, or misshapen.

Since there are either two or six of each component, you can check voltages around all of them and see if one component is out of line with the others. Don't touch anything! We are dealing with high voltages. If a tool slips, you may experience a severe arc flash.

With the unit still powered up, measure the DC bus voltage. In a 480-volt, three-phase system, the DC voltage is higher than the line voltage. It should read 679 VDC. The reason the rectified voltage on the DC bus is higher than the AC line voltage is that a full-wave rectifier's output relates to the AC input's peak-to-peak voltage, not the RMS voltage.

Besides a uniform DC voltage on the DC bus, we want to make sure that there is not appreciable AC ripple. It can be detected by examining the DC bus voltage with either a hand-held, battery-powered oscilloscope or a multimeter set to the AC volts mode. This lowers the instrument's range so that the relatively low ripple voltage can be seen. If an oscilloscope is used, remember that both + and − DC voltage are referenced to and float above ground potential, so a standard grounded bench-type oscilloscope cannot be used unless it is equipped with differential probes. Most technicians use a hand-held, battery-powered oscilloscope with inputs insulated from ground.

Testing Components

The following tests are performed with the VFD in a powered-down state. Power should be shut down and locked out at the main disconnect. We must emphasize that due to the presence of large electrolytic capacitors and the possibility of distributed capacitance, the VFD can harbor lethal voltages long after it is powered down. The proper procedure is to bleed these voltages down by shunting the capacitors and circuitry at appropriate locations using low-ohm, high-watt power resistors equipped with alligator clips. Any capacitor-resistor combination has a time constant, which is dependent upon the resistance and capacitance ratings, so you have to watch the voltage while it slowly declines to a safe level.

Never discharge these electrolytic capacitors by shorting them out with a screwdriver. The current surge will damage the fragile electrolytic layers. Even after the circuitry is discharged, there is no need to touch leads or terminals. As redundant precautions, wear high-voltage gloves and use tools with insulated handles.

The diodes, inductors, and capacitors can be taken out of the circuit for testing by disconnecting one of the two leads. The IGBTs must have two of the three leads disconnected. To test a diode, measure resistance, then reverse the meter's leads and measure resistance again. One way, the ohmmeter battery forward biases the diode, and the other way it reverse biases the diode. Therefore, you will see a significant difference between the two readings. Remember, however, that this is not a dynamic test. A diode may test positive but not perform as expected under operating conditions. However, this test will definitely identify a large proportion of bad diodes.

The two inductors should read identical low resistances, indicating that they are neither open nor shorted. Some inductors read OK when they are cold, but in operation the heat opens up an invisible fissure, breaking the circuit.

The two electrolytic capacitors form a voltage divider with a ground connection tapped at their midpoint and with the positive and negative DC bus lines equidistant from ground. This is why the DC voltages are referenced to but float above ground potential. Connecting a multimeter in ohms mode individually to these capacitors, with one lead of each lifted so that they are temporarily out of the circuit, they will be seen to exhibit a strange behavior. Upon initial connection, the capacitor, depending upon its previous state, will either charge or discharge. Reading ohms, quickly at first, then more slowly, the ohms value will drift smoothly at a speed that depends upon the device's capacitance and the voltage of the meter's internal battery. Eventually a state of equilibrium is reached. Then, if you reverse the leads, the numbers will roll in the opposite direction. Electricians call this strange behavior "counting." Again, this is not a true dynamic test, but it shows that the capacitor is neither shorted nor open, and that it has capacitance.

The IGBTs, as can be seen in the block diagram, Figure 3-6, contain what we may call virtual diodes. There are two in each device, and they can be tested with an ohmmeter. This is only a preliminary test, but it can detect bad semiconductors. For diodes, capacitors, and transistors, there are various types of far more sophisticated testers, including laboratory-grade instruments that inject biases and signals, and compare the output to manufacturers' data sheets.

Of course, in a thorough VFD/motor evaluation, it is equally important to conduct measurements on the output side of the VFD. Low-voltage control signals from the motion controller are applied to the IGBT base terminals in the inverter section, and they modulate the high-voltage from the DC bus. These switching transistors are configured so as to create a high-voltage, high-current, pulse-modulated three-phase supply for the elevator motor. To review, this varying duty cycle electrical energy is capable of varying, in a highly controlled fashion, the speed of the motor without causing it to overheat. The voltage does not vary and therefore the motor can be slowed down or speeded up beyond its rated RPM, and still run efficiently as long

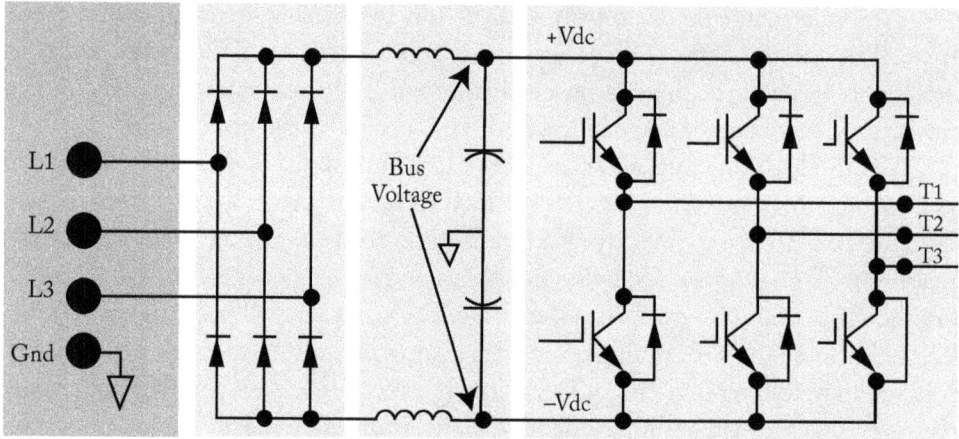

FIGURE 3-6 VFD block diagram.

as bearing and cooling issues are addressed. (In running slower than rated RPM, the internal cooling fan will be turning slower than intended, and supplemental cooling may be needed.)

The pulse-width modulated power that is applied to the motor terminals consists of quasi-square waves, with very fast rise and fall times, in essence high-frequency components. A conventional multimeter cannot accurately measure this non-sinusoidal electrical energy. A true RMS voltmeter measures the heating effect of the pulse-width modulated power, and that is not what we want. The problem with the conventional multimeter is that it has too much bandwidth, responding to high-frequency harmonics, which are not relevant to the motor operation. Also, special shielding is needed to keep out the drive's high-frequency noise.

The Fluke 87V multimeter is well-suited for VFD measurements. It has an internal user-selectable low-pass filter, which can be activated by pressing a yellow button on the front panel. The meter then makes accurate voltage, current, and frequency readings at the VFD output or motor terminals.

At this location and using the specialized meter, voltage readings among the three phases should not vary more than one percent and current readings should not vary more than ten percent.

To determine motor speed, activate the low-pass filter and take a frequency reading between any two phases. Press the Hz button, and the frequency reading will be the motor speed.

So far, we have used a multimeter in volts and ohms mode together with a clamp-on ammeter to check VFD metrics and thereby locate faults. There are other instruments that permit more detailed measurements. While the multimeter is widely and successfully used, there are situations where more advanced instrumentation is warranted.

The oscilloscope is similar to a voltmeter in that it has probes that are used to check voltage potentials between pairs of terminals or conductors in electrical circuitry. The difference is that in addition to displaying metrics such as voltages and frequencies, the oscilloscope is equipped with a screen (originally cathode ray tube, now more user-friendly flat screen) that displays a graph of the waveform. In the most basic time domain configuration, a periodic oscillating waveform is shown with amplitude in volts or millivolts relative to the vertical Y-axis and time in seconds or milliseconds relative to the horizontal X-axis. Thus, the familiar 50 or 60-cycle utility voltage is a sine function, shown in Figure 3-7, and a nine-volt DC signal, shown in Figure 3-8, is a flat horizontal line parallel to the X-axis and intersecting the Y-axis at the nine-volt level.

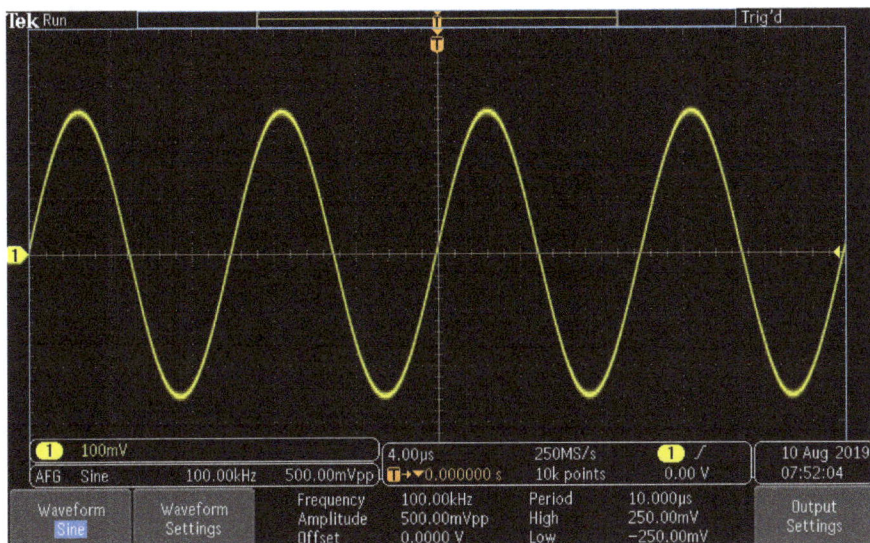

FIGURE 3-7 Sine wave.

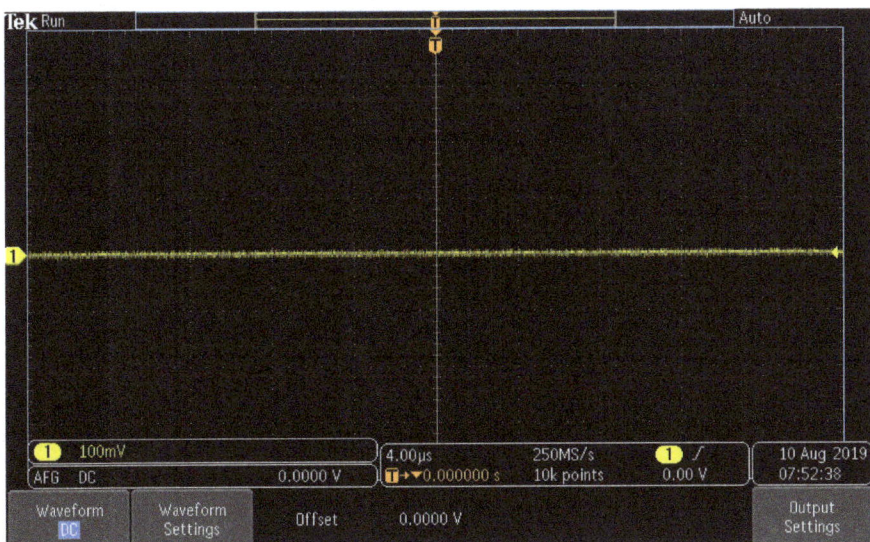

FIGURE 3.8 DC.

The waveform of an electrical signal is based on the source voltage, but the current waveform is greatly modified by the load. This is why looking at waveforms as displayed in an oscilloscope is very useful in troubleshooting an elevator motion controller, VFD, and motor. You can begin at the utility supply, take readings at various points in the VFD, and check the motor, as well as low-amplitude signals throughout the motion controller.

Before we proceed, it is necessary to look once again at a very basic hazard in oscilloscope work. The probe consists of the pointed needle tip or clip-on hook adapter and, in addition, a short wire emanating from the probe body and equipped with an alligator clip at the far end. This is known as the ground return lead. Both of these probe components have to connect to the circuit under investigation in order to extract a signal. The probe tip can be touched to any energized wire or terminal (carefully observing voltage and available current limitations). But in a conventional bench-type, AC-powered oscilloscope, the ground return lead must never contact any voltage that is referenced to but floats above ground potential. When this is done, even if the probe tip is not touching anything, and even if a plugged-in oscilloscope is powered down, there will be a heavy arc-fault involving the energized equipment under investigation, ground return lead, grounded oscilloscope internal metal chassis, facility wiring back to the service entrance panel neutral bar, and utility aerial or underground neutral conductor. Equipment under investigation and oscilloscope may be damaged and the technician injured.

One remedy is to float the oscilloscope without a chassis ground, such as by sawing off the ground prong in the power plug. But this expedient is extremely hazardous because down the road rather than tripping the over-current device, metal parts on the outside of the oscilloscope, such as the analog inputs and exposed hardware on the back panel, can become and remain energized, waiting for someone to touch them.

A three-phase VFD, as well as a switching power supply, has many points where there are voltages referenced to but floating above ground potential, so the grounded bench-type oscilloscope must not be used in this application unless it is equipped with a differential probe set.

The differential probe set is expensive. It reduces the oscilloscope's bandwidth and some models require a separate power supply, so it is not frequently used in VFD work.

A more reasonable alternative is offered by most oscilloscope manufacturers. It is the portable hand-held, battery-powered oscilloscope, which is not connected to the facility ground. This instrument has inputs that are isolated from ground. Most, but not all, have multiple inputs that are insulated from one another, so this point should be verified in the instrument specifications.

The hand-held, battery-powered oscilloscope with isolated inputs is the best way to go in VFD work, and because of its portability, it is very suitable for all-around elevator work.

So far, we have focused on the primary function of the VFD as used in many industrial applications. But in addition to speed, it is fully able to regulate the motor's torque, acceleration, and direction of rotation. Without the VFD, in an AC induction motor, only direction of rotation can be readily controlled, by switching any two of the three-phase connections. The motion controller by means of the VFD inverter section modifies the motor operation in multiple ways as we shall see.

It is logical to start troubleshooting a VFD and motor by using a multimeter to take readings at key locations. If, for example, the elevator has shut down and the motion controller alpha-numeric readout displays an error code that indicates an under-voltage condition, this can certainly be verified by taking various multimeter readings. But an oscilloscope is better able to pinpoint the exact problem: drive, motor, or distorted supply voltage. An oscilloscope will determine if it is a matter of noise, distortion, or transient failures.

First, use the oscilloscope to look at the supply voltage waveforms for each phase at the VFD input. A good supply voltage will consist of well-formed sine waves of equal amplitude for all three phases. The waveforms should not exhibit clipping (flattening at the peaks). You would not want to see electrical energy that is not related to the three fundamental sine waves. This could take the form of jagged, irregular traces, perhaps intermittent, or sharp spikes rivaling the fundamental sine waves.

Harmonics have many sources, most frequently non-linear loads within or even outside of the facility. They may appear only intermittently or at certain times of day, when the offending machinery is running.

Low-frequency harmonics may appear as small sine waves riding on the fundamental. Higher-frequency harmonics appear as an unwarranted thickening or blurring of the fundamental trace. A better determination of the amplitude and spectral location of these harmonics can be made by switching the oscilloscope to its frequency-domain mode.

Here, the X-axis represents frequency rather than time, so the harmonics are shown as spikes separate from the fundamental and diminishing in amplitude as they occur farther in frequency from the fundamental.

Always (unless you are a musician), harmonics are bad. At the very least they represent wasted power, and they can make the VFD malfunction or cause heating in a three-phase neutral conductor or in the motor.

The prominent irregular line at the bottom of the frequency-domain display is the noise floor of the instrument, caused by inevitable thermal noise in internal conductors and components. It is present in all instrumentation and is not a cause for concern.

Previously, we measured the voltage at the DC bus, to determine if it was normal (679 VDC for a three-phase, 480-volt system). Then, we switched the multimeter to AC volts to detect ripple. A more accurate determination can be made using the oscilloscope. Be sure to switch it to AC coupling so that the range is suitable for detecting this small voltage that is riding on the far higher DC voltage.

If there is appreciable ripple, you may want to take a closer look at the capacitors and inductors in the rectifier section. If you replace a capacitor, be sure it has an adequate working volts rating. Watch out for stored voltage!

Now we'll look at the pulse-width modulated output at both the VFD output and at the motor terminals. What we are looking for is harmful reflections, caused by cabling faults and terminations. The problem may be severe due to the fast rise and fall times of the pulses and also the overall frequency of the pulses. The amplitude of the reflected pulses may be caused by cable length, motor load, surge impedances of cable and motor, and rise time, spacing, and magnitude of the drive pulses.

Viewed in the oscilloscope, a normal output waveform at the motor terminals appears as a simple quasi-square wave. If there are reflections, it is likely that you will see the basic square wave with additional lighter extensions above and below the positive- and negative-going waveforms, with distinct boundaries between the basic waveform and the reflected energy.

Assuming the system had been working, the solution should be easy to find. If the VFD was relocated relative to the motor, a different cable may have been substituted, changing the characteristic impedance. If the motor maximum speed was increased, that could give rise to reflections. Or, they could be a consequence of aging motor insulation, in which case it may be a good time to rebuild the motor. Another possible cause could be cable terminations, in which case it is a simple matter to clean and retorque them. Finally, the cable could have been damaged. When conveying high-frequency pulses, the cable may be sensitive to slight imperfections, such as a pinching conduit connector.

STUDY QUESTIONS

1. Most modern elevators are powered by:
   A. nuclear fusion
   B. three-phase induction motors
   C. DC motors
   D. oxen in turnstiles

2. Early signs of motor failure are:
   A. failure to start
   B. reduced power
   C. tripping out
   D. any of the above

3. In the elevator machine room, which of the following is prohibited?
   A. the drive motor
   B. the motion controller
   C. the electric disconnect
   D. Romex wiring

4. It is normal in a DC motor:
   A. for brushes to slowly wear
   B. for the housing to become so hot that you can't touch it
   C. for replacement to be necessary annually
   D. for oil to accumulate underneath on the floor

5. The commutator in a DC motor:
   A. should be replaced annually
   B. never needs replacement
   C. should be inspected periodically
   D. should last the life of the motor

6. Bearing failure may be caused by:
   A. insufficient lubrication
   B. over-lubrication
   C. bearing current
   D. any of the above

7. A cause of bearing current is:
   A. connection to a VFD
   B. misalignment of the motor
   C. running at low speed
   D. ambient heat

8. Bearing current can be reduced by:
    A. running the motor at high speed for one hour
    B. installing CoolBlue common-mode chokes
    C. spray painting the shaft
    D. replacing the bearings

9. Vibration analysis should not be done too often because:
    A. it can crack the motor housing
    B. it is time-consuming and expensive
    C. the motor has to be powered down
    D. none of the above

10. Excessive vibration indicates:
    A. an imbalance
    B. misalignment
    C. incipient bearing failure
    D. any of the above

For answers, go to Appendix A.

CHAPTER 4

ADVANCED MOTOR REPAIR

Elevator diagnostics is a great vocation. Experts in the field are deeply appreciated by building owners and workers, and by office personnel, high-rise residents, and the general public. These people can get to their apartments and offices without climbing stairs. Vertical transportation is incredibly safe. The ride up to office space on the fiftieth floor is less eventful than crossing the street in front of the building. Elevator users are aware of this, and they greatly appreciate the conscientious work performed by the technicians who keep this complex machinery in top condition.

The work is well paid. The only problem is that for some, entry into the field is problematic. A good start is to become a proficient maintenance electrician. If the buildings you are charged with maintaining have elevators, and if you demonstrate an aptitude and interest in elevator maintenance, you are well positioned to learn the trade.

The fact is, however, that there are distinct levels of expertise in this field. It is important to progress in an orderly fashion, obtaining the required licensing and certifications as you go along. You must absolutely not get in over your head and you must avoid making a mistake that will expose users to injury or worse.

What It Takes

This complex work is highly stratified in terms of knowledge, expertise, and the tasks that can be safely undertaken. When an elevator system goes down, a novice worker can check doors and perhaps reset the motion controller, but many operations call for greater experience and knowledge of advanced diagnostic instrumentation and tooling. To perform a job like regrooving a sheave, outside firms that specialize in such tasks must be called in.

One such organization is Renown Electric, based in Canada. They work throughout North America, rebuilding, rewinding, and servicing electric motors. They specialize in elevator maintenance and repair, providing in-shop and field work on a 24-7 basis. On their website, Renown-electric.com, you will find a great amount of technical information, including e-books, archived white papers, and videos documenting many aspects of elevator maintenance, diagnostics, and repair. I am grateful to this outstanding organization for the many photographs and abundant technical information provided for this book.

When it comes to elevator motor teardown and advanced repair, especially rewinding, most technicians send the motor to a specialized shop. There the work is done by experts under highly-controlled conditions with equipment such as metal-turning lathes, dipping tanks, and large ovens for removing moisture. This is done at higher temperatures for burning off old insulation prior to rewinding.

Some operations performed in specialized motor shops are bearing changes, frame repair, machining end bells, machining shafts, milling keyways, dynamic balancing, and dielectric testing.

Some Examples

Elevator technicians may not have access to the facilities, expertise, and tooling required to perform all of these tasks, but having a good understanding about what is involved provides valuable perspective and insight regarding a wide range of problems from a diagnostic point of view. Naturally, there is some overlap. For example, in a DC motor, brush replacement is straightforward and can usually be performed while power is locked out, without disconnecting the motor from the driven equipment. Alternatively, this is done in a motor shop as a final operation in a total rebuild.

In rebuilding a DC motor, the armature should be rewound using Class B to Class H insulation. The goal is to provide protection for the armature and field coils so that they can withstand higher internal heat rise and not undergo premature insulation failure. In an overheating situation, actually the motor is not always at fault. If possible, improve air flow in the machine room, and check for binding in the load and poor quality in the electrical supply.

Major motor repair work generally begins with Megger and surge tests. These are dielectric tests intended to evaluate windings and insulation integrity, particularly to find out if the motor is shorted to ground in any way. This work may be done in the field, with the motor mounted in place. We will discuss the process in detail, first stressing that because it involves applying high voltage to the motor windings for specific lengths of time, great care must be taken to ensure that the motor is disconnected from all external circuitry. This is done so that workers are not exposed to the hazardous voltages, and that outside equipment and components are not damaged, and also that the test is not invalidated because terminals are shunted and perhaps grounded.

There are numerous diagnostic tools that are helpful in ascertaining the exact cause of motor failure or poor performance. Multimeters, clamp-on ammeters, temperature sensors, vibration analyzers, Meggers (shown in Figure 4-1), winding analyzers, oscilloscopes, and spectrum analyzers are basic instruments that are used in motor diagnosis.

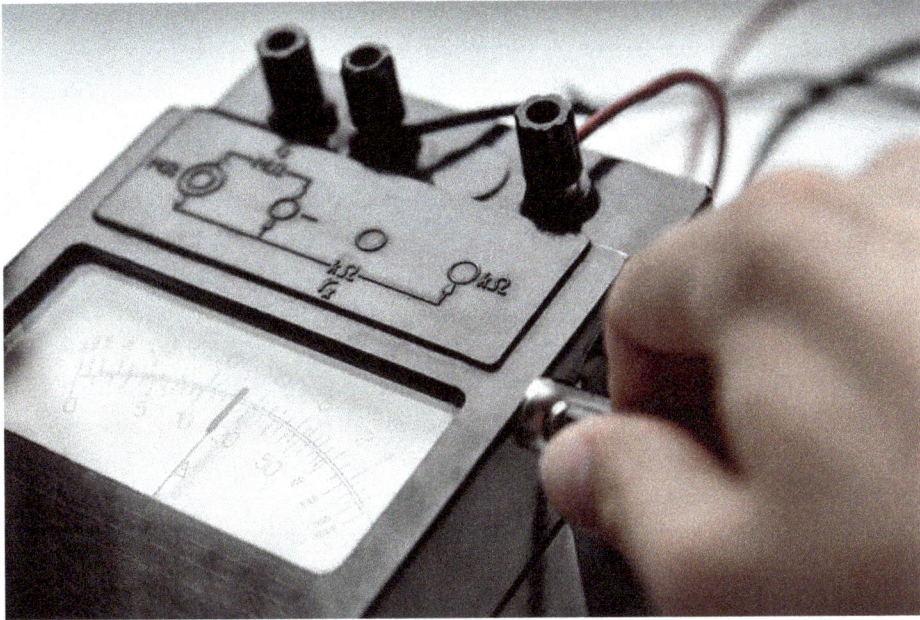

FIGURE 4-1 The Megger (trade name) is an instrument that is used to measure resistance and evaluate electrical insulation in motors, electrical equipment, and cable.

Dielectric testing using a Megger, as opposed to resistance tests with the multimeter in ohms mode, should be performed if the initial visual inspection, voltage, current, and ohmmeter tests are not definitive. A Megger test, as it is commonly known, consists of applying high AC or DC voltage between electrical conductors and the motor frame. To assess the integrity of new winding insulation, the usual test consists of applying 1000 volts plus twice the motor voltage for 60 seconds.

The reason this test is effective is that when a high voltage is applied to a non-conductor, the electrons attempt to create an ionized path. That is what happens in the atmosphere when a lightning bolt occurs. The purpose of the Megger test is to detect thin or weak spots in the winding insulation. Of course, this is a sort of war of nerves, because you never know when you are weakening the insulation in the course of testing it, so that the next time current could break through. Megger testing should be done with a light touch.

Additionally, care must be taken if the technician is to avoid shock hazard. The hand-crank, dynamo Megger is safer than the plug-in type, because as soon as a shock is felt, the individual stops cranking, whereas the convulsive effect of sustained electrical current on muscle tissue can prevent the worker from releasing an energized object. To prevent overstressing the insulation, subsequent tests are run at 85 percent of full strength and for reconditioned insulation, 60 percent.

Stator insulation is a vulnerable part of any motor. Insulation can be compromised by heavy, long-term use, stator motion at startup caused by strong magnetic fields, manufacturing defects, and chemical deposits or contamination from over-greasing.

IEEE 522 contains information on setting voltage limits for various conditions. Megger test results are expressed in resistance. Periodic tests as part of a preventive program are valuable for keeping track of motor winding resistance over time, so that preventive action can be taken before an outage occurs. Once again, we'll emphasize that any motor undergoing Megger testing must be electrically disconnected and kept away from any conductive objects during the course of the procedure.

Megohmmeter high-voltage insulation resistance testers were first used in the 1880s as high-voltage power distribution and end-use equipment became available. The Megger is used to verify insulation integrity not only in a motor, but wherever there are conductors that convey high voltage or require exceptional reliability.

It is inevitable that insulation quality with respect to resistance degrades in time, particularly under adverse conditions such as temperature, humidity, moisture, and dust. Repeated electrical and mechanical stress also negatively impact insulation integrity. Megger testing detects the resulting current leakage paths by injecting high voltage by means of probes to the outside of insulated cables. Never touch both probes to bare metal. Since heat and water are prime offenders in insulation degradation, Megger tests are often performed on electrical wiring and equipment where a fire has been successfully suppressed without total loss.

There are three types of Meggers in common use today:

- **Electronic type, battery-operated.** This type is quite accurate and very portable. With a digital display, it is easy to read while recording data.
- **Hand-operated.** The user turns a crank to power the internal dynamo. It is safest, because in case of shock the user invariably stops cranking.
- **Motor-operated.** An external power source rotates a motor-generator set within the instrument. This provides a high degree of isolation between electrical input and output.

Grainger.com and other online outlets offer a wide range of Megger insulation resistance testers including an excellent hand-cranked model with analog readout for about $1500.

As emphasized in Chapter 3, a large motor should be inspected regularly and any changes in temperature or sound should be noted and investigated. Thermal imaging and vibration testing are very useful. If a motor fails to start, runs intermittently, produces excessive heat, or trips out, it's time to look for the cause. Often the problem is external to the motor—the load is jammed, binding, or mismatched. Check the guides in the hoistway. Depending on the manufacturer's recommendations, they may have to be lubricated periodically, but you need to determine the correct lubricant and method of application that is compatible with the safeties.

Assuming that external causes including abnormal loading, usage, and poor power quality have been eliminated, it is time to bring out the test equipment and enter into a totally open-minded information-gathering mode.

The Hipot test for dielectric strength, as previously noted, involves applying electrical energy between internal motor circuits and the grounded frame. Before beginning, plan the test based on the electrical rating of the motor and its time in service. The motor power supply is to be locked out, the motor electrically isolated from external wiring and no persons close to the motor; when assessing new windings, the standard test is run by continuously applying 1000 volts, 50 or 60 Hz, plus two times the motor rated voltage for 60 seconds.

The Hipot test should be conducted at the full strength only once, and subsequent tests should be done at 85 percent of full strength. Repeated full-strength tests can over-stress the insulation. If the insulation has been reconditioned, run tests at 60 percent of full strength.

Surge Testing

The surge test uses different equipment. It is useful in detecting motor burnout as well as potential for future failure. Surge testing, shown in Figure 4-2, evaluates shorts between individual windings before the problem becomes more widespread. These faults can be caused by chemical deposits and over-greasing, manufacturing or rewinding errors, startup movement of the stator, and excessively heavy use.

FIGURE 4-2   Surge testing consists of applying voltage to individual sets of motor windings for the purpose of identifying partial failure before a total motor rebuild becomes necessary.

Using a Baker or Electrom testing machine, technicians can safely apply a voltage pulse (surge) to each set of motor windings to isolate its performance individually and in comparison to one another. Then, remedial measures can be planned.

IEEE 522 provides standards for surge testing including appropriate voltage levels for many winding types and conditions.

Using an ordinary multimeter in the ohms mode, this is a very useful test that ascertains the quality and efficiency of the motor circuit under test.

A basic load is probed and the meter measures the voltage drop at various points. Because electric current follows the path of least resistance, in the event of a fault the excess current will create a voltage drop reading. If the circuit has been broken, the meter can create a temporary flow of current to isolate the problematic area.

An indication of voltage drop is often an early warning sign that cleaning, maintenance, or routine repair is needed.

Core-Loss Test

The core-loss test indicates the difference between a motor's input and output power. All motors exhibit core loss, but if it is excessive, that is a sign of excess heat and trouble down the road. A core loss tester measures this parameter. A sign of significant core loss can reveal problems at an early stage so that they can be repaired before widespread damage occurs.

The above tests are performed in accordance with ANSI/EASA Standard AR100-2105, which outlines best practices for the testing and repair of rotating electrical motors.

Detecting and Analyzing Motor Vibration

In an electric elevator motor, or for that matter in any machine with rotating parts, freedom from vibration contributes to a long service life and reduces the chance of a harmful or even catastrophic outage. A smooth rotating shaft, good bearings, and just the right amount of lubrication contribute to smooth operation. Another favorable condition is uniform and symmetrical mass distribution. If the rotor is out of balance, it is sure to vibrate, especially at a certain critical speed. Perfect symmetry can never be achieved, but we must strive toward this ideal situation insofar as possible.

Excessive vibration causes noise and damages all moving parts, especially the bearings. Smooth and energy-efficient operation is achieved by static and dynamic balancing.

Static unbalance should be eliminated first. It consists of a relatively heavy angular fraction of the rotating member when at rest. It is present and measurable when the rotor is not turning. Placed on a smooth surface, an out of balance rotor will roll, eventually coming to rest with the heavy spot at the bottom. A rotor that is substantially in a state of static balance will continue to roll until it is slowed by friction and stops, not necessarily always in the same angular position. Static unbalance can be corrected by adding or subtracting weight at one or more points on the perimeter.

In a world that is not ideal, a rotor will have an infinite number of unbalances, however small, distributed along the axis of rotation. These unbalances can be simplified as two unbalances in separate planes relative to the center. These unbalances have differing magnitudes and angular locations.

Dynamic balancing consists of rotating the body and causing it to vibrate, analyzing the vibration, and taking corrective action by adding or removing weights so that the resultant mass becomes aligned with the rotational axis. A condition for achieving dynamic balance is a satisfactory state of static balance. It is a two-stage process. This is done in an automotive tire shop by rotating the mounted tire and fastening small lead weights as needed to the tire rim.

Bearing Maintenance

Bearing problems are a frequent contributing factor in motor failure. Motor damage usually occurs before bearing wear is noticed. Early detection is essential. The best tools are a thermal imager and mechanic's stethoscope, which can be fabricated from a piece of flexible tubing.

Correct installation of replacement bearings and the right amount of lubrication are critical. Insufficient lubrication leads to rapid bearing failure. Many maintenance workers are unaware that over-lubrication also sets the stage for short bearing life. When too much grease has been injected into the bearing, the motor strains to churn the lubricant. Abnormal friction and pressure heat the lubricant, which degrades it. When excess grease is expelled and accumulates inside the motor enclosure, it inevitably finds its way into the windings, degrading insulation.

An even more common, if less known, cause of premature bearing wear is shaft voltage, which manifests as bearing current. We see this mostly in induction motors driven by variable frequency drives, but it occurs also in less frequently-used DC and synchronous AC motors.

It may seem unlikely that VFDs, shown in Figure 4-3, could play a role in premature bearing wear and failure, but here's how it happens:

A VFD uses direct current from its internal DC bus, creating a variable duty-cycle quasi-square wave. A low-voltage digital signal from the motion controller is applied to insulated gate bipolar transistors (IGBTs). These switching devices (six in a three-phase motor) supply high-voltage, high-current pulses to power the motor.

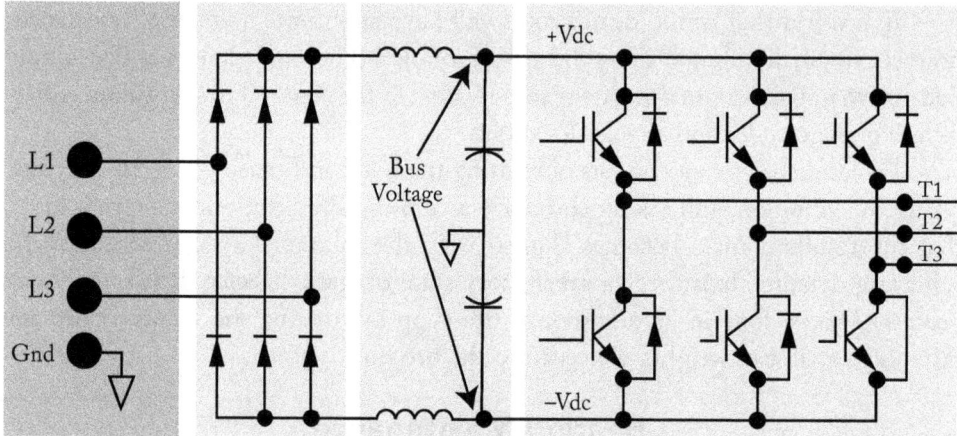

FIGURE 4-3 Variable frequency drive block diagram, showing six IGBTs in output section. The high-speed pulsed output to the motor may cause bearing current, wear, and eventual failure.

The high-frequency pulses have very fast rise and fall times, and the pulses fire at variable rates. Consequently, they cross insulating barriers within the motor because of low capacitive reactance, moving as current through the bearing to the motor housing to ground.

This discharge causes premature and cumulative wear consisting of fluting, scoring, cracks, and fractures. An early symptom is frosting of the otherwise polished metal surfaces. Also, the grease darkens due to an accumulation of burnt particles.

Bearing current can take the following forms, resulting in early bearing failure:

- Low-frequency circulating currents follow the path of least resistance. They do not leave the shaft and go to ground. This common-mode current is induced in the air gap between stator and rotor and follows a path through the motor frame and motor bearings, bypassing the shaft.
- High-frequency shaft grounding currents pass through the shaft and go to ground via driven equipment such as an elevator hoist.
- Capacitive discharge currents cross the air gap between stator and rotor, pass through the bearing, and go to ground through the shaft.

A mechanic's stethoscope can detect a worn bearing before the motor is damaged. Both bearings will probably need to be replaced, and to protect the new ones the bearing current should be eliminated. First detect and measure the bearing current to determine if that is actually the problem.

Shaft voltage is only present while the motor is running, particularly at a high speed where the pulse-width modulated frequency is higher. This may be difficult to measure because the spikes are very brief. A high-bandwidth instrument equipped with a carbon-brush probe is needed to contact the rotating shaft. The Fluke ScopeMeter, shown in Figure 4-4, with AEGIS shaft voltage probe is preferable to a digital multimeter in this application.

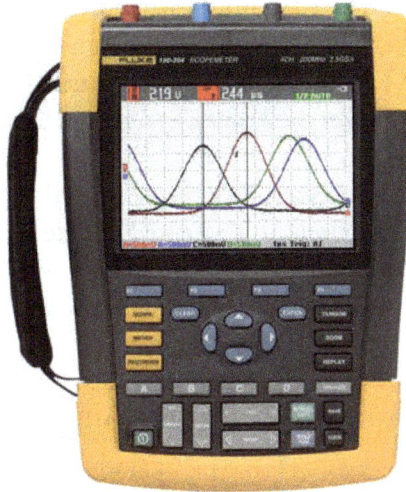

FIGURE 4-4 The Fluke ScopeMeter equipped with a shaft probe is perfect for measuring shaft voltage in a VFD/induction motor. (*Fluke*)

The high bandwidth and fast sampling rate will detect the presence of shaft voltage and determine whether it is high enough to cause arcing through the bearing grease. This equipment is not required to measure the first of the three types of bearing current listed above, because it does not pass through the shaft.

Motors powered by a utility sine wave usually exhibit one or two volts at the shaft. When a VFD with rapidly switching pulses enters the picture, you may measure 8 to 15 volts, which is enough to arc through grease and cause early bearing failure.

Measuring shaft voltage on a powerful motor while it is running can be problematic if not dangerous. The VPS420-R shaft voltage probe interfaces with the shaft, connecting to the meter by means of a small conductive carbon brush that rides on the spinning shaft. For this measurement, clip the ground return lead to the motor's enclosure. Also, an i400s current probe can be clamped over one of the conductors running from VFD to motor.

The Fluke ScopeMeter is capable of Connect-And-View automatic triggering, which displays any relevant signal, and ScopeRecord, which stores waveforms in the instrument's memory so that they can be saved, analyzed, and sent to colleagues. The ScopeMeter's isolated channels permit the user to view simultaneously current and voltage.

By way of preventive maintenance, there are non-intrusive tests that can be performed, often with the motor running and in service. Related to examination by stethoscope is vibration spectrum analysis. A steady vibration modulated by periodic spikes, for example, is characteristic of fluting in a bearing.

When symptoms appear, the bearings should be replaced and shaft voltage and bearing current can be eliminated by creating a jumper around the bearing. This jumper should be between the motor enclosure, which is at ground potential, and the shaft. The problem with this method is that the shaft is rotating, so contact has to be made by means of a conductive brush, which is subject to wear and could fail. The other method is by installing CoolBlue common-mode chokes, also known as inductive absorbers. They last indefinitely and are easily installed by disconnecting power cables, slipping the chokes over them, and re-attaching cables. (For details, see Coolblue-mhw.com.)

In a manufacturing facility, run to failure, as opposed to preventive and predictive maintenance, is not a viable option. It is essential to detect potential mechanical and electrical problems before they occur, thereby avoiding an outage. Reliable motor operation is essential. The following sections detail things you can do to ensure trouble-free operation.

More on Vibration Analysis

Vibration analysis, shown in Figure 4-5, is particularly useful in detecting stator faults, which otherwise cannot be detected without motor teardown. In a large motor, especially when starting or decelerating rapidly, the stator coils are greatly stressed. Their mounts can deform, permitting sufficient motion of the stator so that it contacts the rotor, damaging both sections. Also, considerable heat rise can be present.

Field vibration analysis can be performed using a portable instrument that identifies vibrations and displays their waveforms, showing frequency and amplitude.

The Fluke 810 Vibration Tester checks rotating machinery for unbalance, looseness, misalignment, and bearing failure. It works for motor rotational speeds of 200 to 12,000 RPM. Diagnosis details include plain text readout, fault severity (slight, moderate, serious, and extreme), repair details, cited peaks, and spectra.

FIGURE 4-5   Fluke 810 Vibration Tester. (*Fluke*)

The instrument is auto-ranging. It has four channels, 24 bit. The usable bandwidth is 2.5 to 50 kHz. The dynamic range is 128 dB with 100dB signal-to-noise ratio. The FFT resolution is 100. Non-volatile memory consists of SD memory card, 2 GB internal, plus a user-accessible slot for additional storage.

The price is just under $10,000.

Many DC elevator motors with diligent brush and commutator maintenance no doubt are capable of providing satisfactory operation with long intervals between total overhauls. The trend, however, has been to upgrade to VFD-induction motor systems, and where the building owners are willing to capitalize the changeover, there is much to be said for this option.

Nikola Tesla's three-phase induction motor was a brilliant innovation. Rather than feeding heavy current into the spinning rotor through brushes and a commutator, in an induction motor the stator and rotor are essentially the primary and secondary of a transformer, and by means of induction power is transferred. No more brushes! The only moving part is the spinning rotor, supported by bearings. With adequate lubrication, protected from heat, vibration, and bearing current, the induction motor will last a good long time.

Similarly, gearless drives are replacing the old geared systems, and here again there is less maintenance and greater efficiency. There is the expense of changing over, but the benefits are significant. One problem with a geared drive is that even with timely oil changes, there is bound to be friction and heat. Crown and worm gears have finite life spans. As they wear, minute metal particles compromise the lubricant, and wear accelerates. The only thing you can do is to carefully monitor vibration and temperature.

Gearless systems are relatively trouble free. Modern gearless systems driven by VFD-powered induction motors can last many years, especially if ropes and sheaves are inspected and serviced as needed. Moreover, due to their smaller size and space requirements, gearless systems are easier to repair and replace, entailing less downtime. There is no need for oil changes, with their risks of oil leaks and related catastrophes.

Additionally, in gearless designs there is the potential for Leadership in Energy and Environmental Design (LEEDS) certification recognizing enhanced energy efficiency. Tell your building owners about tax credits and reduced insurance premiums. Quieter operation and higher car speeds are additional benefits.

Elevator brakes work in conjunction with the motor, and these two systems should be inspected and maintained so that they work in harmony. It is important that the brakes, governed by the motion controller, in normal operation are applied only after the car has stopped moving. In other words, they should not be fighting the motor as that would lead to rapid wear. On the other hand, they should be robust and capable of stopping the car in an emergency situation such as unexpected car movement, power outage, or free fall.

Since the brakes are such a vital part of overall elevator operation, they should be inspected regularly as part of a comprehensive maintenance program. Dirt and dust accumulation traps and holds moisture, and this type of contamination is bad for electrical contacts as well as for steel linkages and mechanical subsystems. Inexperienced maintenance workers tend to over-lubricate to compensate for uncertain intervals, and this spells trouble. If a bearing is over-lubricated, it will spin harder as it churns excess grease, and soon it will run hot until the grease expands and some of it is expelled, damaging the seals in the process. Picking up dirt and moisture, the grease will find its way to electrical insulation and terminations, brake pads, metal surfaces, and other sensitive locations. Other than not over-lubricating in the first place, the answer is to keep floors, machinery, and other surfaces clean and dry. Any sign of oil leakage should be investigated and repaired immediately.

If brake components show signs of premature wear, the motion controller may be applying the brakes before the car has stopped, or not releasing them before it starts. It may be necessary to contact the manufacturer. Improper brake maintenance inevitably results in loss of brake pad material, which can be seen in a large air gap between disc and coil carrier; the end result is brake dragging and/or delayed brake release. The first step in brake inspection is to ascertain the air gap between disc and brake pad. Depending on the size of this air gap, you can decide whether adjustments will suffice or new parts will be needed.

Use a feeler gauge. Check with the manufacturer to determine acceptable tolerances. The following are typical:

If the air gap is greater than 0.02 inches and less than 0.04 inches, the brakes can be adjusted by loosening the lock screws and turning the setting bolts. All setting bolts are adjusted in this manner and the locking screws are retightened. If the original air gap was greater than 0.04 inches, replacement pads and possibly other parts are needed. Lock screws may have Allen heads or they may be metric or inch-type bolts. In any event, a torque wrench is needed to tighten them to specifications contained in the manufacturer's documentation. Beware of overtightening as this can wear the threads, setting the stage for failure down the road.

Hydraulic Rope Equalizers

Traction elevators invariably have several wire ropes that work in concert and provide redundancy in the interest of efficient operation and safety. The car goes up and down the length of the hoistway many times per day, often with a heavy load, day after day and year after year. In the fullness of time there is bound to be rope and sheave wear. Worn sheave grooves cause accelerated rope wear, and worn ropes cause accelerated sheave wear, so regular inspection and timely maintenance are essential.

Worn ropes must be replaced, but worn sheaves can be re-machined if the damage is detected in time. It is the responsibility of the elevator maintenance technician to develop a sharp eye and report the first sign of wear so that it can be rectified without extensive down time or excessive cost. Obviously, ropes and sheaves last far longer if there is equal tension on all of them.

Human technicians can adjust individual rope tensions, manually estimating the lateral deflection by touch. But the advantage in hydraulic rope equalizers is that they act automatically, instantaneously, and whether the car is moving or not. In addition to saving on rope and sheave costs, equalized tension allows the traction drive and entire elevator system to operate more quietly and efficiently. Cable life is increased close to 50 percent.

The ropes interact with sheaves by wrapping around them at the motor drive, top and bottom of the hoistway, and at counterweights. In all cases, due to their tension the ropes are pressed tightly into grooves that are milled into the sheaves and pulleys. Hydraulic rope equalizers maintain equal pressure between ropes and grooves so that re-machining is needed less frequently. Hydraulic rope balancing benefits users by providing a smoother and potentially faster ride. Machinery cost is reduced by eliminating the need for springs and associated hardware; if a rope stretches beyond the capacity of the hydraulic equalizers to maintain uniform tension, a limit switch generates a trouble alarm.

A critical factor in the efficiency and safety in the operation of a traction elevator is the fit between steel ropes and sheaves. As emphasized above, equal tension on all ropes is of great importance. Additionally, ropes must maintain a firm grip on the sheaves. Increased traffic and heavier loads result in accelerated wear in these metal-on-metal components. While traditionally sheave regrooving may have been required once in the service life of an elevator system, currently more frequent inspection is required. Space in new buildings is increasingly at a premium, and this has meant smaller machine rooms and even machine room-less designs. Smaller components are being used, with reduced circumference sheaves and tighter rope bends. The traditional U-groove sheave can be upgraded to Undercut-U and Progressive-V designs, which increase traction dramatically by improving gripping action.

With the help of sheave groove evaluation kits such as those provided by Renown Electric, in-house maintenance technicians can evaluate the condition of sheave grooves at appropriate intervals. Renown Electric can perform on-site regrooving or, when necessary, re-machining of the sheaves at their facilities. If needed, bearings are replaced. Performed in a timely fashion, this work reduces the possibility of an unexpected interruption in elevator service.

Passenger Rescue from a Stuck Car

Electrical power to an elevator motor and brake can be interrupted for a variety of reasons ranging from a utility-wide outage to a command from the motion controller. If there is a power outage, the local emergency power supply should start, come up to speed, and be online within a few seconds. The transfer switch will energize preselected essential loads, including the elevators. This, of course, may not go as planned. Also, there are other events that may cause an elevator to go out of service. When power is interrupted, the motor stops and the brakes are applied. The bottom line is that the car stops abruptly, rather than proceeding to its destination or even as far as the closest landing.

Removing occupants from a car that is stuck in this fashion is known in the trade as doing an extraction. Some elevator servicing companies strongly advise on-site maintenance personnel to wait for the outside specialists to arrive. The problem here is that in a general outage, these full-time elevator technicians may not be available for hours.

The fire department may be able to help, especially if in the area they cover there are many buildings including high-rises with fast traction elevators and low-rise buildings with hydraulic elevators. On-site workers, however, are first to respond and if they have the expertise, should be able to complete the extraction. Building and maintenance department managers should establish in advance procedures for extracting passengers from a stuck elevator car. Several factors need to be considered:

- Is there a medical emergency?
- Is there a fire or other hazardous event?
- What are the mental states and wishes of the stranded passengers?
- What is the level of expertise of on-site personnel?
- How long is the interruption in service expected to last?
- If outside help is coming, what is the estimated time of arrival?

Procedures for extracting passengers from a stuck elevator car can vary, depending upon the type and age of the elevator system, and the position of the car relative to landings. If the car happens to have stopped within a few inches of a landing, it is a simple task to manually open the hoistway and car doors and to assist the passengers as they step out of the car, up or down into the lobby.

With the hoistway door open, due to safety interlocks, it is impossible for the car to move. However, in the context of an elevator malfunction, redundant safeguards are advised. Before beginning the extraction, the best course of action is to open and lock out the disconnect so that the motor cannot unexpectedly receive electrical power. The disconnect is in a medium-sized enclosure mounted on the wall within sight of the motion controller and the motor. This disconnect can take the form of a three-pole (for a three-phase system) breaker, which can be manually moved to the off position. It will never automatically reset. Alternatively, the disconnect may consist of three large cartridge fuses in an enclosure with an external handle that in the down position shuts off the power. The main disconnect should be clearly labeled on its front cover. If there is any doubt, you can follow the large (2" or so) electrical conduit from outside the machine room to the disconnect, then to the motion controller and motor. Only the motor uses this kind of power.

Electricians have small padlocks that are designed for locking out disconnects, and one of these should be used to ensure that the motor is not powered up while the extraction is in progress. Additionally, a worker should remain in the machine room during the extraction to ensure that because of a breakdown in communication, other technicians do not attempt to restart the elevator by disabling the disconnect lockout and jumping out the door interlock. With the power supply locked out, the motor will remain powered down and the drive brake fully engaged so that the car cannot move.

The machine room, in compliance with applicable codes, is to be kept locked to prevent unauthorized entry. Keys may be kept in the facility front office or maintenance office. When attempting to restart the elevator, the door is often temporarily kept from closing. This is one reason for having a person present in the machine room for the duration of the extraction.

When the power is out, it is not feasible to move a stuck car in a traction elevator. However, in a hydraulic elevator this may be possible by opening a valve to bleed just enough hydraulic fluid from the cylinder back into the reservoir so that the car descends until it is aligned with the landing floor, greatly simplifying the extraction.

Once assured that the elevator will not restart, technicians can open the hoistway door, which is locked as always when the car is not at the landing. Near the top of the hoistway door on the lobby side is a small, inconspicuous hole, accessible by means of a stepladder. Into this hole must be inserted a peculiar hinged drop key, which must be seen to be appreciated. The articulated section drops down to unlock the door, which should be opened very carefully and the position of the car observed. Care must be taken. The greatest number of elevator fatalities occur when a worker steps through an open hoistway door into a darkened shaft.

After the hoistway door has been opened, it is time to open the car door. Depending on the type of elevator, this door can be opened with a moderate amount of force while manipulating the door linkage on the car top. The procedure may be covered in the operations manual, and it should be investigated and rehearsed in advance.

If the correct hoistway door has been opened and the car floor is within a few inches of the lobby floor, the passengers can simply step out. Sometimes there is a greater gap and a stepladder is required, either on the inside of the car to get up to the next higher landing, or on the outside to get down to the next lower landing.

Passengers who are being extracted range from totally terrified to amused and entertained by the whole process. They are reassured if one of the maintenance workers enters the car and assists from that end. Fire rescue persons have a harness and tether that are used to move reluctant, hysterical, or otherwise helpless individuals. If there is a medical emergency, the problem is compounded and an EMT or doctor may be needed. Generally, people do what is necessary.

Traveling Cable

In an elevator installation, the traveling cable is a necessary link between the motion controller and the elevator car. Most of the run consists of fixed wiring secured to and supported by building framing in the conventional manner. However, eventually you come to the hoistway, and because the car follows a vertical path moving a distance substantially equal to the height of the building, extraordinary demands are made on this traveling cable. In design and installation, the cable has to last at least 20 years with as many as three million flex cycles.

Low-rise, slower moving elevators may use the less expensive flat cable, sometimes consisting of several separate parallel segments. More commonly, particularly in high-rise, faster elevators, traveling cable consists of a round profile assembly of numerous very finely stranded copper conductors arranged around a central steel supporting member. These wires are covered by a specialized jacket that combines flame and abrasion resistance and good low-temperature performance. Polyvinyl chloride is a frequently-used thermoplastic that is cost effective and suitable for many installations. In demanding environments, advanced polymers are specified. An outer polyurethane coating provides enhanced abrasion resistance where needed. In some areas, halogenated materials are not permitted, and costly polyolefin compounds are used.

The traveling cable is located in the elevator hoistway. One end is terminated on the bottom of the car and the other end is terminated at the top of the hoistway. At no point does the traveling cable touch the floor of the pit. It hangs in the hoistway between the points of attachment, and depending upon the location of the car within the hoistway, the entire weight of the cable is supported by either one termination or the other.

The traveling cable conveys all power and signal information to and from the car. It is not a great amount of power—just enough to open and close the car door, provide a moderate amount of lighting and, where required, small heating and/or cooling loads. As for signaling, there are call buttons, alarm, telephone, door interlocks, car levelling, and similar low-current demands. Overall, the electrical requirements are not severe, but the traveling cable has some physical challenges that must be confronted. If failure occurs, there could be an interruption of service while the entire cable, substantial in a high-rise, is replaced. A 100-story building, for example, may in a heavy wind sway as much as three feet in ten-second periods. Given the fact that the traveling cable has no small inertia of rest, there is a certain tendency for it to contact the hoistway walls, not to mention counterweights and compensating chains.

The traveling cable must be terminated at both ends so that it cannot pull free, which would damage it beyond repair. It can be terminated using a self-tightening device or looping it around hardware that is attached to the hoistway and fastening the end of the cable to itself. The most used terminating method consists of inserting the traveling cable into a clamping fixture in the opposite direction from which the pulling force is exerted. That way the cable does not have to be bent and tools are not required. When tension is applied by the weight of the hanging traveling cable, the fixture tightens like a vise, holding the cable securely so it is not in danger of coming loose.

In evaluating methods for terminating the traveling cable, the first consideration is always to avoid tight loops. Severe bends of the traveling cable stress the conductors. The concern is not only that they will work harden and break due to cyclic flexing, but also that minute fractures will reduce the ampacity of the copper conductors so that in those that carry power, hot spots will develop, leading to failure. In signaling conductors, similar damage can cause data corruption and interfere with communication so that the phone link becomes noisy.

As loading fluctuates with changing locations of the car within the hoistway, the traveling cable tends to untwist under load and retwist as the loading decreases. We mentioned the unwanted relative lateral motion of the traveling cable when the building sways due to heavy winds. A related problem is cable vibration during car motion. In severe cases, the cable can contact hoistway walls, wire ropes, and compensating chains. When this motion or vibration is repeated many times, there is an increased exposure to physical damage.

Traveling cables are carefully engineered to reduce the possibility of this sort of damage. Among the techniques that have succeeded is the selection of low torsion wire rope for the steel support member. Also, alternating the stranding direction of adjacent layers in round traveling cable has been effective. Proper twist rate with respect to cable diameter further minimizes exposure to damage.

Flat traveling cables must also be designed so that damage from many usage cycles is minimized. A valuable technique in flat cable design has been to alternate direction of twist in adjacent conductors.

With enhanced elevator performance, especially in terms of car speed and hoistway height, new traveling cable designs have been installed. The approximately 30 conductors in older traveling cables tended to be a uniform size, 16 AWG. Contemporary traveling cables contain a far greater number of conductors, sometimes as many as 120. They are variously sized for specific applications. Current-carrying conductors are 14 AWG. Signaling conductors are 18 AWG. Communications conductors are 20 AWG, shielded. And there may be a coaxial cable for closed-circuit TV. A recent innovation is the use of optical fiber, which is highly flexible and immune to electromagnetic interference.

Lubrication and Oil Analysis

Moving metal parts that are in contact experience friction. Consequences are temperature rise and wear at the areas of contact. A film of oil between the moving parts enables them to slide freely. Water is a good lubricant for some materials such as rubber on glass or metal, but where metal parts are in contact, oil or grease is needed.

Elevator systems are comprised of numerous components where there is metal-to-metal contact of moving parts. In some of these, lubrication is not appropriate. For example, where the metal rope in a traction elevator wraps about a grooved sheave, friction is essential and any form of lubrication is undesirable. Similarly, car door opening and closing mechanisms generally run dry. A dab of grease at linkage joints may seem useful, but it would quickly attract abrasive material, increase wear, and accelerate failure. Where the door slides in a grooved sill, lubrication is not advised since dirt would likewise accumulate here. The best maintenance procedure in this area is regular use of a vacuum cleaner to keep the grooves free of foreign material.

Elevator manufacturers recommend regular lubrication of the hoistway guide rails. A proprietary lubricant can be applied using a paint brush or directly from a spray can. The manufacturer's documentation must be consulted to ensure that this will not interfere with the operation of the safeties, which are attached to the car and grip the guide rails in a free-fall situation.

Mechanical drive assemblies that consist of meshing gears normally run in an enclosed gearbox that is filled to a prescribed level with petroleum-based or synthetic oil. Unlike in an automotive engine, there is no combustion, so the oil change intervals are less frequent. They should nevertheless be closely observed since eventually the oil will break down and become a less efficient lubricant. Input and output shaft seals eventually experience wear and loosen, allowing loss of oil, as shown by a wet appearance on the outside of the seal and oil stains on the floor. Gear oil is quite heavy and resists leakage. Also, it clings to the bearing surfaces, so it is good in that way. On the other hand, for higher speed and more complex machinery, a lighter oil may be specified. If there are rubber seals that may deteriorate, automatic transmission fluid is advised. In all cases, adhere to the manufacturer's recommendations regarding oil type and oil change intervals.

Elevator systems, with the exception of outside construction elevators, are not ordinarily powered by internal combustion engines. However, elevator technicians may be tasked with maintaining the backup power generators. Buildings large enough to have an elevator system generally have ones that are diesel-powered. Invariably, such engines have extensive operating manuals, which should be consulted regarding maintenance procedures. The main areas of concern are checking and changing oil as needed, fuel and air filters, fan belts, starting system including battery charging, and fuel supply.

Besides a temperature gauge on the gearbox, it is a good idea to take a look with a thermal imager. Any unexplained temperature rise should be investigated. Periodic oil analysis is a great predicter of potential problems down the road. Technicians can determine the presence of excessive metal particles, contamination consisting of dirt or moisture, and inevitable breakdown and loss of lubricating quality. Oil samples can be taken at scheduled intervals and sent to laboratories for comprehensive analysis. To receive a valid readout, definite procedures need to be followed when collecting the sample. The motor or gearbox should be run until normal operating temperature is attained. This ensures that any particulate contamination will not be sitting at the bottom of the gearbox, and thus will not be included in the sample.

Normally the sample is extracted by removing the filler plug and inserting a flexible tube with squeeze bulb attached. To obtain an accurate sample, the end of the tube should be midway between top and bottom of the fluid. Important characteristics of the sampled fluid include viscosity, water or coolant contamination, fuel dilution, sludge content, and use of the wrong lubricant.

Comparison of the sample with records from previous samplings will determine whether there is a damaging trend that cannot be resolved by a simple oil change. This information will determine appropriate remedial action, leading to a range of benefits including prevention of bearing failure, avoidance of an elevator outage, and extended product life.

STUDY QUESTIONS

1. In rebuilding a DC motor:
   - A. it is best to use Class B to Class H insulation
   - B. begin by hosing down the old motor
   - C. begin by doing electrical tests on the old motor
   - D. it's not necessary to install new brushes

2. In doing a Megger test:
   - A. the motor should be powered up and running
   - B. you are determining if a winding is grounded
   - C. the instrument should be left powered up overnight
   - D. 1000 volts plus five times the motor voltage is applied for ten minutes

3. In an electric motor, stator insulation can be compromised by:
   - A. long, heavy use
   - B. stator motion at start-up
   - C. chemical deposits
   - D. any of the above

4. Megger tests were first conducted:
   - A. in the 1880s
   - B. around 1900
   - C. during World War II
   - D. after 2000

5. The surge test:
   - A. will not predict motor failure
   - B. finds faults between individual windings
   - C. will not detect rewinding errors
   - D. all of the above are true

6. The core-loss test indicates:
   - A. the difference in a motor between input and output power
   - B. motor vibration
   - C. motor temperature
   - D. winding resistance to ground

7. Freedom from vibration:
   - A. results solely from correct lubrication
   - B. shows that brushes are good
   - C. indicates symmetrical mass distribution
   - D. is no longer considered important

8. Static imbalance:
   A. should be measured first
   B. results from a relatively heavy fraction of the rotating member when at rest
   C. is present when the motor is not turning
   D. all of the above

9. Dynamic balancing:
   A. consists of rotating the body and causing it to vibrate
   B. should be measured before static balancing is attempted
   C. cannot be done in the field
   D. is hazardous due to potential for electric shock

10. Driving machine brake pads can be adjusted if the air gap is:
   A. less than .02 inches
   B. .02 to .04 inches
   C. .05 inches
   D. .06 inches

For answers, go to Appendix A.

CHAPTER 5

TROUBLESHOOTING ELEVATOR SYSTEMS

The American Society of Mechanical Engineers (ASME), through its Board of Safety Codes and Standards (BSCS), develops and maintains a comprehensive portfolio of codes and standards that govern, among many other things, elevators. A comprehensive listing can be seen at ASME.org. ASME A17, Safety Code for Existing Elevators and Escalators, is an essential reference for elevator maintenance workers and repair technicians. It can be ordered online.

As an introduction to our discussion of troubleshooting elevator systems, we repeat because of their relevance here some pertinent provisions in ASME A17, beginning at Section 3.8, Driving Machines, and Sheaves, shown in Figure 5-1.

Section 3.8.1, General Requirements, provides that sheaves and drums are to be made of cast iron or steel with finished grooves for ropes. (Throughout, when we talk of ropes, we are referring to steel ropes, as defined in this standard.) Set screw fastenings are not to be used in lieu of keys or pins on connections subject to torque or tension. Friction gearing or a clutch mechanism is not to be used to connect a driving-machine drum or sheave to the main driving mechanism, other than in connection with a car leveling device.

Section 3.8.2, Winding Drum Machines states that these machines are to be provided with a slack-rope device having an enclosed switch of the manually reset type that will cause the electric power to be removed from the elevator driving-machine motor and brake if the hoisting ropes become slack or broken.

Final terminal stopping devices for winding drum machines are to consist of a stopping switch located on the driving machine and a stopping switch located in the hoistway and operated by cams attached to the car.

Stopping switches, located on and operated by the driving machine, are not to be driven by chains, ropes, or belts. The opening of these contacts is to occur before or coincident with the opening of the final terminal stopping switch.

FIGURE 5-1 The sheave must be cast iron or steel. (*Renown Electric*)

Where a three-phase AC driving-machine motor is used, as in Figure 5-2, the mainline circuit to the driving-machine motor and the circuit of the driving-machine brake coil are to be directly opened either by the contacts of the machine stop switch or by stopping switches mounted in the hoistway and operated by a cam attached to the car.

FIGURE 5-2 Current to a three-phase AC driving motor is to be interrupted by stopping switches mounted in the hoistway and operated by a cam attached to the car. (*Judith Howcroft*)

Driving machines equipped with a DC motor and DC brake are permitted to have the final terminal stopping device contacts installed in the operating circuits. The occurrence of a single ground or the failure of any single magnetically operated switch, contactor, or relay must not render any final terminal stopping device ineffective.

Section 3.8.4, Brakes, provides that the elevator driving machine is to be equipped with a friction brake applied by a spring or springs, or by gravity and released electrically. The brake is to have a capacity sufficient to hold the car at rest with its rated load. For passenger elevators and freight elevators permitted to carry employees, the brake is to be designed to hold the car at rest with an additional load up to 25 percent in excess of the rated load.

Section 3.9, Terminal Stopping Devices states that enclosed upper and lower normal stopping devices are to be provided and arranged to slow down and stop the car automatically at or near the top and bottom terminal landings. Such devices are to function independently of the operation of the normal stopping means and of the final terminal stopping device.

Normal stopping devices are to be located on the car, in the hoistway, or in the machine room, and are to be operated by the movement of the car.

Broken rope, tape, or chain switches are to be provided in connection with normal stopping devices located in the machine room of traction elevators. Such switches are to be opened by a failure of the rope, tape, or chain and are to cause the electrical power to be removed from the driving machine motor and brake.

Section 3.9.2, Final Terminal Stopping Devices, states that enclosed upper and lower final terminal electromechanical stopping devices are to be provided and arranged to prevent movement of the car by the normal operating devices in either direction of travel after the car has passed a terminal landing. Final terminal stopping devices are to be located as follows:

Elevators having winding drum machines are to have stopping switches on the machines and also in the hoistway operated by the movement of the car.

Elevators having traction driving machines are to have stopping switches in the hoistway operated by the movement of the car.

Section 3.10.4, Electrical Protective Devices, states that these devices are to be provided as follows:

a. **Slack-Rope Switch.** Winding drum machines are to be provided with a slack-rope device equipped with a slack rope switch of the enclosed manually reset type that will cause the electric power to be removed from the elevator driving-machine motor and brake if the suspension ropes become slack.

b. **Motor-Generator Running Switch.** Where generator-field control is used, means are to be provided to prevent the application of power to the elevator driving machine motor and brake unless the motor generator set connections are properly switched for the running condition of the elevator. It is not required that the electrical connections between the elevator driving-machine motor and the generator be opened in order to remove power from the elevator motor.

c. **Compensating Rope Sheave Switch.** Compensating rope sheaves are to be provided with a compensating rope sheave switch or switches mechanically opened by the compensating rope sheave before the sheave reaches its upper or lower limit of travel to cause the electric power to be removed from the elevator driving-machine motor and brake.

d. **Broken Rope, Tape, or Chain Switches Used in Connection with Machine Room Normal Stopping Switches.** Broken rope, tape, or chain switches are to be provided in connection with normal terminal stopping devices located in machine rooms of traction elevators. Such switches are to be opened by a failure of the rope, tape, or chain.

e. **Stop Switch on Top of Car.** A stop switch is to be provided on the top of every elevator car, which will cause the electric power to be removed from the elevator driving-machine motor and brake. It is to be of the manually operated and enclosed type with red operating handles or buttons. It is to be conspicuously and permanently marked "STOP" and indicate the stop and run positions. It is to be positively opened mechanically, opening not solely dependent on springs.

f. **Car Safety Mechanism Switch.** A switch is required where a car safety is provided.

g. **Speed Governor Overspeed Switch.** Where required by Section 3.6.1, a speed governor overspeed switch is to be provided.

h. **Final Terminal Stopping Devices.** For every elevator, final terminal stopping devices are to be provided.

i. **Emergency Terminal Speed Limiting Device.** Where reduced stroke oil buffers are provided, emergency terminal speed limiting devices are required.

j. **Motor-Generator Overspeed Protection.** Means are to be provided to cause the electric power to be removed automatically from the elevator driving-machine motor and brake, should a motor-generator overspeed excessively.

k. **Motor Field Sensing Means.** Where DC is supplied to an armature and shunt field of an elevator driving-machine motor, a motor field current sensing means is to be provided, which will cause the electric power to be removed from the motor armature and brake unless current is flowing in the shunt field of the motor.

A motor field current sensing means is not required for static control elevators provided with a device to detect an overspeed condition prior to, and independent of, the operation of the governor overspeed switch. This device must cause power to be removed from the elevator driving-machine motor armature and machine brake.

l. **Buffer Switches for Oil Buffers Used with Type C Car Safeties.** Oil level and compression switches are to be provided for all oil buffers used with Type C safeties.

m. **Hoistway-Door Interlocks or Hoistway-Door Electric Contacts.** Hoistway-door interlocks or hoistway-door electric contacts are to be provided for all elevators.

n. **Car Door or Gate Electric Contacts.** Car doors, as shown in Figure 5-3, or gate electric contacts are to be provided for all elevators.

FIGURE 5-3 All doors are to have electric contacts. (*Wikipedia*)

o. **Normal Terminal Stopping Devices.** Normal terminal stopping devices are to be provided for all elevators.

p. **Car Side Emergency Exit Electric Contact.** An electric contact is to be provided on every car side emergency exit door.

q. **Electric Contacts for Hinged Car Platform Sills.** Where provided, hinged car platform sills are to be provided with electric contacts.

r. **In-Car Stop Switch.** On passenger elevators equipped with non-perforated enclosures, a stop switch, either key operated or behind a locked cover, is permitted to be provided in the car and located in or adjacent to the car operating panel. The switch is to be clearly and permanently marked "STOP" and is to indicate the stop and run positions. The switch is to be positively opened mechanically and its opening is not to be solely dependent on springs. When opened, this switch is to cause the electric power to be removed from the elevator driving-machine motor and brake.

s. Emergency Stop Switch. On all freight elevators, passenger elevators with perforated enclosures, and passenger elevators with non-perforated enclosures not provided with an in-car stop switch, an emergency stop switch is to be provided in the car and located in or adjacent to each car operating panel. When open ("STOP" position), this switch is to cause the electric power to be removed from the elevator driving-machine motor and brake and is to:

1. be of the manually operated and closed type;
2. have red operating handles or buttons;
3. be conspicuously and permanently marked "STOP" and indicate the stop and run positions; and,
4. have contacts that are positively opened mechanically (opening not solely dependent on springs).

t. Stop Switch in Pit. A stop switch conforming to (e) (Stop Switch on Top of Car) is to be provided in the pit of every elevator. The switch is to be adjacent to every pit access.

u. Buffer Switches for Gas Spring Return Oil Buffers. A buffer switch is to be provided for gas spring return oil buffers that will cause electric power to be removed from the elevator driving-machine motor and brake if the plunger is not within 0.5 inches of the fully extended position.

Section 3.10.5, Power Supply Disconnecting Means, provides that:

a. A disconnect switch, as shown in Figure 5-4, or a circuit breaker is to be installed and connected into the power supply line to each elevator motor or motor generator set and controller. The power supply line is to be provided with over-current protection, preferably inside the machine room.

b. The disconnect switch or circuit breaker is to be of the manually closed multi-pole type, and be visible from the elevator driving machine or motor generator set. When the disconnecting means is not within sight of the driving machine, the control panel, or the motor generator set, an additional manually operated switch is to be installed adjacent to the remote equipment and connected to the control circuit to prevent starting.

c. No provision is to be made to close the disconnect switch from any other part of the building.

d. Where there is more than one driving machine in a machine room, disconnect switches or circuit breakers are to be numbered to correspond to the number of the driving machine that they control.

FIGURE 5-4 Disconnect removes power from motor so that it stops and from brake so it is applied. (*Judith Howcroft*)

Section 3.10.6, Phase Reversal and Failure Protection, states that elevators having polyphase AC power supply are to be provided with means to prevent the starting of the elevator motor if the phase rotation is in the wrong direction or if there is a failure of any phase.

This protection is considered to be provided in the case of generator field control having AC motor-generator driving motors, provided a reversal of phase will not cause the elevator driving-machine motor to operate in the wrong direction. Controllers on which switches are operated by polyphase torque motors provide inherent protection against phase reversal or failure.

Section 3.10.7, Operating of the Driving Machine with a Hoistway Door Unlocked or a Hoistway Door or Car Door Not in the Closed Position. This is permitted under the following conditions:

a. By a car-leveling or truck zoning device

b. When a hoistway access switch is operated

c. When the top-of-car or in-car inspection operation utilizing a car-door bypass of hoistway-door bypass switch is activated

Devices other than those specified above are not to be provided to render hoistway-door interlocks, the electric contacts of hoistway-door combination mechanical locks and electric contacts, or car door, gate electric contacts, or car door or gate inlocks inoperative. Existing devices that do not conform to the above are to be removed.

Section 3.10.8, Release and Application of Driving-Machine Brakes, states that driving-machine brakes are not to be electrically released until power has been applied to the driving-machine motor.

Two devices are to be provided to remove power independently from the brake. If the brake circuit is ungrounded, all power lines to the brake are to be opened. The brake is to apply automatically when:

a. The operating device of a car switch or continuous pressure operation elevator is in the stop position

b. A floor stop device functions

c. Any of the electrical protective devices in Section 3.10.4 functions

Under conditions described in (a) or (b), the application of the brake is permitted to occur on or before the completion of the slowdown and leveling operations.

The brake is not to be permanently connected across the armature or field of a DC elevator driving motor.

Section 3.10.9, Control and Operating Circuit Requirements, provides that the failure of any magnetically operated switch, contractor, or relay to release in the intended manner, or the occurrence of a single accidental ground, or combination of accidental grounds, is not to permit the car to start or run if any hoistway door interlock is unlocked or if any hoistway door or car door or gate electric contact is not in the closed position.

Section 3.10.10, Absorption of Regenerated Power, states that when a power source is used which, in itself, is incapable of absorbing the energy generated by an overhauling load, means for absorbing sufficient energy to prevent the elevator from attaining governor tripping speed or a speed in excess of 125 percent of rated speed, whichever is lesser, is to be provided on the load side of each elevator power supply line disconnecting means.

Section 3.10.12, System to Monitor and Prevent Automatic Operation of the Elevator with Faulty Door Contact Circuits, states that means are to be provided to monitor the position of power-operated car doors that are mechanically coupled with the landing doors while the car is in the landing zone in order to:

a. Prevent automatic operation of the car if the car door is not closed, regardless of whether the portion of the circuits incorporating the car door contact or interlock contact of the landing door coupled with the car door, or both, are closed or open, except as permitted in 3.10.7.

b. Prevent the power closing of the doors during automatic operation if the car door is fully open and if any of the following conditions exist:

 1. The car door contact is closed, or a portion of the circuit incorporating this contact is bypassed.
 2. The interlock contact of the landing door that is coupled to the opened car door is closed, or the portion of this circuit incorporating this contact is bypassed.
 3. The car door contact and the interlock contact of the door that is coupled to the opened car door are closed, or the portions of the circuits incorporating these contacts are bypassed.

Section 3.11, Emergency Operation and Signaling Devices, Section 3.11.1, Car Emergency Signaling Devices, states that in all buildings, the elevator(s) are to be provided with the following:

a. If installed, altered, or both under ASME A17.1-2000 or earlier edition:

 1. An audible signaling device, shown in Figure 5-5, operable from the emergency stop switch, when provided, and from a switch marked "ALARM" that is located in or adjacent to each car operating panel. The signaling device is to be located inside the building and audible inside the car and outside the hoistway. One signaling device is permitted to be used for a group of elevators.

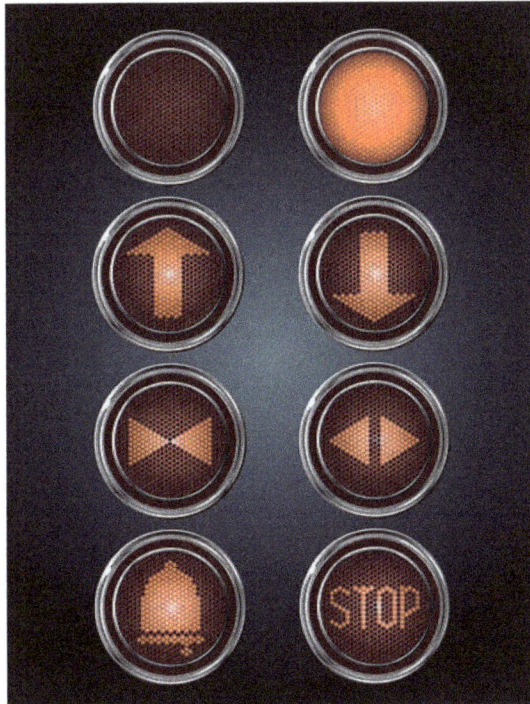

FIGURE 5-5 An alarm operated by a switch in the car. (*Wikipedia*)

2. Means of two-way communication (telephone, intercom, etc.) between the car and a readily accessible point outside the hoistway that is available to emergency personnel. The means to activate the two-way communication system does not have to be provided in the car.

3. If the audible signaling device, or the means of two-way communication, or both, are normally connected to the building power supply, they are to automatically transfer to a source of emergency power within ten seconds after the normal power supply fails. The power source is to be capable of providing for the operation of the audible signaling device for at least one hour, and the means of two-way communication for at least four hours.

4. In buildings in which a building attendant (building employee, watchman, etc.) is not continuously available to take action when the required emergency signal is operated, the elevators are to be provided with a means within the car for communication with or signaling to a service that is capable of taking appropriate action when a building attendant is not available.

5. An emergency power system is to be provided conforming to the requirements of (a) (3).

b. If installed, altered, or both under ASME A17.1a—2000 or later editions, the emergency communications system is to comply with Section 2.27 of the ASME A17.1/CSA b44 Code under which it was installed or altered.

Section 3.11.2, Operations of Elevators Under Standby (Emergency) Power, states that an elevator is permitted to be powered by a standby (emergency) power system provided that, when operating on such standby power, there is conformance to the requirements of Section 3.10.10.

Section 3.11.3, Firefighters' Service, provides that elevators are to conform to the requirements of ASME/ANSI A17.1—1987 Rules 211.3 through 211.8 unless at the time of installation or alteration it was required to comply with a later edition of A17.1.

All elevators that are part of a group are to conform to identical firefighters' service operation requirements regardless of which edition of A17.1 they complied with at the time of their installation or alteration.

The Phase I and Phase II switches for all elevators in a building are to be operable by the same key.

Section 3.12, Suspension Means and the Connections, provides that cars are to be suspended by steel wire ropes attached to the car frame or passing around sheaves attached to the car frame. Only iron (low-carbon steel) or steel wire ropes, having the commercial classification "Elevator Wire Rope," or wire rope specifically constructed for elevator use is to be used for the suspension of elevator cars and for the suspension of counterweights.

Section 3.12.4, Minimum Number and Diameter of Suspension Ropes, provides that all elevators except freight elevators that do not carry passengers or freight handlers and have no means of operation in the car are to conform to the following requirements:

a. The minimum number of hoisting ropes used is to be three for traction elevators and two for drum-type elevators. Where a car counterweight is used, the number of counterweight ropes used is not to be less than two.

b. The minimum diameter of hoisting and counterweight ropes is to be 0.375 inches. Outer wires of the ropes are to be not less than 0.024 inches in diameter.

Section 3.12.5, Suspension Rope Equalizers, states that suspension rope equalizers, where provided, are to be of the individual compression spring type.

Section 3.12.6, Securing of Suspension Ropes to Winding Drums, provides that suspension wire ropes of wind drum machines are to have the drum ends of the ropes secured on the inside of the drum by clamps or by tapered babbitted sockets, or by other approved means.

Section 3.12.7, Spare Rope Turns on Winding Drums, provides that suspension wire ropes of winding drum machines are to have not less than one turn of the rope on the drum when the car is resting on the fully compressed buffers.

Section 3.12.8, Suspension Rope Fastenings, states that spliced eyes by return loop are permitted to continue in service. Suspension rope fastenings are to conform to requirement 2.20.9 of ASME A17.1—2004 when the ropes are replaced.

Section 3.12.9, Auxiliary Rope Fastening Devices, states that auxiliary rope fastening devices, designed to support elevator cars or counterweights if any regular rope fastening fails, are permitted to be provided.

Part IV, Hydraulic Elevators, applies to direct plunger and roped-hydraulic elevators.

Section 4.2.1, Buffers and Bumpers, states that car buffers and bumpers are to be provided. Solid bumpers are to be permitted in lieu of buffers where the rated speed is 50 feet per minute or less. Car frames and platforms, car enclosures and capacity, and loading are to conform to the requirements for traction elevators.

Section 4.3.1, Connection to Driving Machine, provides that the driving member of a direct plunger driving machine is to be attached to the car frame or car platform with fastenings of sufficient strength to support that member. The connection to the driving machine is to be capable of withstanding, without damage, any forces resulting from a plunger stop.

Section 4.3.2, Plunger Stops, states that plungers are to be provided with solid metal stops and/or other means to prevent the plunger from traveling beyond the limits of the cylinder. Stops are to be so designed and constructed as to stop the plunger from maximum speed in the up direction under full pressure without damage to the connection to the driving machine, plunger, plunger connection, couplings, plunger joints, cylinder, cylinder connecting couplings, or any other parts of the hydraulic system. For rated speeds exceeding 100 feet per minute where a solid metal stop is provided, means other than the normal terminal stopping device (i.e., emergency terminal speed limiting device) are to be provided to retard the car to 100 feet per minute with a retardation not greater than gravity before striking the stop.

Section 4.3.3, Hydraulic Elevators, states that hydraulic elevators that have any portion of the cylinder buried in the ground and that do not have a double cylinder or a cylinder with a safety bulkhead are to:

a. Have the cylinder replaced with a double cylinder or a cylinder with a safety bulkhead protected from corrosion by one or more of the following methods:

   1. Monitored cathodic protection
   2. A coating to protect the cylinder from corrosion that will withstand the installation process
   3. A protective plastic casing immune to galvanic or electrolytic action, salt water, and other known underground conditions

Or:

b. Be provided with a device meeting the requirements of Section 3.5 or a device arranged to operate in the down direction at an overspeed not exceeding 125 percent of rated speed. The device is to mechanically act to limit the maximum car speed to the buffer striking speed, or stop the elevator car with rated load with a deceleration not to exceed 32.2 feet per second^2 and is not to automatically reset. Actuation of the device is to cause power to be removed from the pump motor and control valves until manually reset.

Or:

c. Have other means to protect against unintended movement of the car as the result of uncontrolled fluid loss.

Section 4.4.1, Pump Relief Valve, provides:

a. **Pump Relief Valve Required.** Each pump or group of pumps is to be equipped with a relief valve conforming to the following requirements, except as covered by (b):

   1. **Type and Location.** The relief valve is to be located between the pump and the check valve and is to be of such a type and so installed in the bypass connection that the valve cannot be shut off from the hydraulic system.
   2. **Size.** The size of the relief valve and bypass is to be sufficient to pass the maximum rated capacity of the pump without raising the pressure more than 50 percent above the working pressure. Two or more relief valves are permitted to be used to obtain the required capacity.
   3. **Sealing.** Relief valves having exposed pressure adjustments, if used, are to have their means of adjustment sealed after being set to the correct pressure.

b. **Pump Relief Valve Not Required.** No relief valve is required for centrifugal pumps driven by induction motors, provided the shutoff, or maximum pressure which the pump can develop, is not greater than 135 percent of the working pressure at the pump.

Section 4.4.2, Check Valve, states that a check valve is to be provided and so installed that it will hold the car with rated load at any point when the pump stops or the maintained pressure drops below the minimum operating pressure.

Section 4.4.3, Mechanically Controlled Operating Valves, provides that mechanically controlled operating valves are not to be used. Existing terminal stopping devices consisting of an automatic stop valve independent of the normal control valve and operated by the movement of the car as it approaches the terminals, where provided, may be retained.

Section 4.4.4, Supply Pipes and Fittings, states that supply pipes and fittings are to be in sound condition and secured in place.

Section 4.5, Tanks, provides in Section 4.5.1 General Requirements:

a. **Capacity.** All tanks are to be of sufficient capacity to provide for an adequate liquid reserve to prevent the entrance of air or other gas into the system.

b. **Minimum Liquid Level Indicator.** The permissible minimum liquid level is to be clearly indicated.

Section 4.5.2, Pressure Tanks relates to:

a. **Vacuum Relief Valves.** Tanks subject to vacuum sufficient to cause collapse are to be provided with one or more vacuum relief valves with openings of sufficient size to prevent collapse of the tank.

b. **Gage Glasses.** Tanks are to be provided with one or more gage glasses attached directly to the tank and equipped to shut off the liquid automatically in case of failure of the glass. The gage glass or glasses are to be so located as to indicate any level of the liquid between permissible minimum and maximum levels, and are to be equipped with a manual cock at the bottom of the lowest glass.

c. **Pressure Gage.** Tanks are to be provided with a pressure gage that will indicate the pressure correctly to not less than 1½ times the pressure setting of the relief valve. The gage is to be connected to the tank or water column by pipe and fittings with a stop cock in such a manner that it cannot be shut off from the tank except by the stop cock. The stop cock is to have a "T" or lever handle set in line with the direction of flow through the valve when open.

d. **Inspector's Gage Connection.** Tanks are to be provided with a 0.25 inch pipe size valve connection for attaching an inspector's gage while the tank is in service.

e. Liquid Level Detector. Tanks are to be provided with a means to render the elevator inoperative if for any reason the liquid level in the tank falls below the permissible minimum.

f. Handholes and Manholes. Tanks are to be provided with means for internal inspection.

g. Piping and Fittings for Gages. Pipings and fittings for gage glasses, relief valves, and pressure gages are to be of a material that will not be corroded by the liquid used in the tank.

Section 4.6, Terminal Stopping Devices, provides that terminal stopping devices are to conform to the requirements of Section 3.9.1.

Section 4.7.1, Operating Devices, states that operating devices are to conform to the requirements of Sections 3.10.1 and 3.10.2.

Section 4.7.2, Top-of-Car Operating Devices, states that top-of-car operating devices are to be provided and must conform to the requirements of 3.10.3 except for uncounterweighted elevators having a rise of not more than 15 feet.

The bottom normal terminal stopping device is permitted to be made ineffective while the elevator is under the control of the top-of-car operating device.

Section 4.7.3, Anti-creep Leveling Devices, states that each elevator is to be provided with an anti-creep leveling device conforming to the following:

a. It is to maintain the car within three inches of the landing irrespective of the position of the hoistway door.

b. For electrohydraulic elevators, it is to operate the car only in the up direction.

c. For maintained pressure hydraulic elevators, it is required to operate the elevator in both directions.

d. Its operation is permitted to depend on the availability of the electric power supply provided that:

    1. The power supply disconnecting means required by 3.10.5 is kept in the closed position at all times except during maintenance, repairs, and inspections.

    2. The electrical protective devices required by 4.7.4 (b) do not cause the power to be removed from the device.

Section 4.7.4, Electrical Protective Devices, states that electrical protective devices conforming to the requirements of 3.10.4, where they apply to hydraulic elevators, are to be provided and operate as follows:

a. The following devices are to prevent operation of the elevator by the normal operating device and also the movement of the car in response to the anti-creep leveling device:

1. Stop switches in the pit
2. Stop switches on top of the car
3. Car side emergency exit door electric contacts, where such doors are provided

b. The following devices are to prevent the operation of the elevator by the normal operating device, but the anti-creep leveling device required by 4.7.3 is to remain operative:

1. Emergency stop switches in the car
2. Broken rope, tape, or chain switches on normal stopping devices when such devices are located in the machine room or overhead space
3. Hoisting-door interlocks or hoisting-door electric contacts
4. Car door or gate electric contact
5. Hinged car platform sill electric contacts
6. In-car stop switch, where permitted by 3.10.4(t)

Section 4.7.5, Power Supply Line Disconnecting Means, states that power supply line disconnecting means are to conform to the requirements of 3.10.5.

Section 4.7.6, Devices for Making Hoistway-Door Interlocks or Electric Contacts, or Car Door or Gate Electric Contacts Inoperative, states that the installation is to conform to the requirements of 3.10.7.

Section 4.7.7, Control and Operating Circuit Requirements, provides that control and operating circuits are to conform to the requirements of 3.10.9 and 3.10.12.

Section 4.7.8, Emergency Operation and Signaling Devices, states that emergency operation and signaling devices are to conform to the requirements of 3.11.

Section 4.8, Additional Requirements for Counterweighted Hydraulic Elevators, states that counterweighted hydraulic elevators are to be roped so that the counterweight does not strike an overhead structure when the car is resting on its fully-compressed buffer. Counterweighted hydraulic elevators are to conform to the requirements of Section 3.2 where applicable.

Where counterweights are provided, counterweight buffers are not to be provided.

Section 4.9, Additional Requirements for Roped Hydraulic Elevators. Section 4.9.1, Top Car Clearance, provides that roped-hydraulic driving machines, whether of the vertical or horizontal type, are to be so constructed and so roped that the piston will be stopped before the car can be drawn into the overhead work. The top car clearance is to meet the requirements of 2.4.4.

Section 4.9.2, Top Counterweight Clearance and Bottom Counterweight Runby, states that where a counterweight is provided, the top clearance and bottom runby are to conform to the following:

a. **Top Clearance.** The top clearance is not to be less than the sum of the following:

1. The bottom car runby

    2. The stroke of the car buffers used

    3. Six inches

  b. Bottom Runby. The bottom runby is not to be less than the sum of the following:

    1. The distance the car can travel above its top terminal landing until the piston strikes its mechanical stop

    2. Six inches

    The minimum runby specified is not to be reduced by rope stretch.

Section 4.9.3, Protection of Spaces Below Hoistway, provides that where the hoistway does not extend to the lowest floor, the space below the pit is to be enclosed with permanent walls or partitions to prevent access.

Section 4.9.4, Piston Stops, states that piston stops are to be provided to bring the piston to rest at either end of the piston travel from maximum speed in the up direction, under full pressure without damage to the driving machine, piston, piston joints, cylinder, cylinder couplings, or any other part of the hydraulic system.

For rated speeds of over 100 feet per minute where a solid metal stop is provided, means other than the normal terminal stopping device are to be provided to retard the car to 100 feet per minute with a retardation not greater than gravity, before striking the stop.

Section 4.9.5, Piston Connections, states that:

  a. Equalizing Crosshead. Where more than one piston is used on puller-type roped hydraulic elevators, an equalizing crosshead is to be provided for the attachment of the rods to the traveling sheave frame to ensure an equal distribution of the load to each rod.

  b. Equalizing or Cup Washers. Equalizing or cup washers are to be provided under piston rod nuts to ensure a true bearing.

  c. Piston Rods. Piston rods of puller-type hydraulic elevators are to have a safety factor of no less than eight based on the cross-sectional area at the root of the thread of the material used. A true bearing is to be maintained under the nuts of both ends of the piston rod to prevent eccentric loadings on the rod.

Section 4.9.6, Car Safety Devices. Car safety devices conforming to the requirements of Section 3.5, except 3.5.2, are to be provided. Counterweight safeties are not to be provided.

Section 4.9.7, Speed Governors, states that speed governors conforming to the requirements of Section 3.6 are to be provided.

Section 4.9.8, Sheaves, states that sheaves are to be cast iron or steel and are to have finished grooves for ropes.

The traveling sheaves are to be guided by means of metal guides and guide shoes. The guide shoes are permitted to be equipped with nonmetallic inserts. Sheave frames, where used, are to be constructed of structural or forged steel and are to be designed and constructed with a factor of safety not less than eight for the material used.

Single continuous straps (known as U-strap connection) are not to be used for frames or as connections between piston rods and traveling sheaves.

Section 4.9.9, Slack-Rope Device, states that roped-hydraulic elevators are to be provided with a slack-rope device and switch of the enclosed, manually reset type that will cause the electric power to be removed from the motor and the valves if the hoisting ropes become slack or are broken.

Section 4.9.10, Suspension Ropes and Their Connections, provides that all elevators except freight elevators that do not carry passengers or freight handlers and have no means of operation in the car are to conform to the following requirements:

a. Suspension ropes are to conform to the requirements of 3.12.1 through 3.12.3, 3.12.5, 3.12.8, and 3.12.9.

b. The minimum number of hoisting or counterweight ropes used for roped hydraulic elevators is to be not less than two.

c. The minimum diameter is to be 0.375 inches, and the outer wires of the rope are not to be less than 0.024 inches in diameter.

The purpose in going through these elevator construction mandates is to demonstrate the great many protective interlocks and safety features that have evolved over the last couple of centuries. Typically, devices were embedded in the design to prevent various catastrophes, such as a broken hoisting rope that would allow a free-falling car to fall to the bottom of the hoistway and crash in the pit. Elisha Otis's safeties consisted of clamps fastened to the car that would engage the guides anchored to the inside of the hoistway walls. Sensing a slack-rope condition, they would stop the car and hold it in place.

This illustrates the process: a dangerous condition is sensed and action is taken prior to any catastrophic event that might cause loss of life, injury, or damage to property. Today, everything is electrical. Throughout ASME A17, the following wording or variations occur:

"When the switch is opened, electric power is to be removed from the elevator driving-machine motor and brake."

Depending on gearing and friction in the drive, when electric power is removed from the driving motor, the car will stop, especially if it is counterweighted. In case the weight of the car could spin the de-energized motor, the brake would provide redundant protection. Good design provides that electric power, rather than engaging the brake, is required to hold it open. If this power fails or is removed, in addition to the motor stopping, the brake will engage. Naturally, the car will stop even if between floors and filled with passengers.

If the elevator is to be out of service for any length of time, the passengers will have to be allowed to exit the car. This is known as an "extraction." Elevator service firms insist that only their trained personnel, not facility maintenance workers, should do a passenger extraction. This may not be realistic due to availability of those technicians and the distance they may have to travel. Leaving the passengers confined to the car for an extended period carries its own health and other risks. In any event, a pre-established policy should be in place, and if appropriate, facility workers should be familiar with the extraction procedure.

A specialized key will open the hoistway door from outside. This key should be kept in a secure location. It is generally inserted at an inconspicuous location at the top of the hoistway door. As a further precaution, it has to be inserted in a unique way if the hoistway door is to be opened. If the car is stopped between floors, better access is usually available from either the floor above or the floor below the car location. Often a stepladder must be passed into the car so that the passengers can exit. During this procedure, it is prudent to lock out the elevator power at the disconnect in the machine room. Additionally, the car should not be able to move while the hoistway door is open. In the event that a passenger extraction becomes necessary, fail-safe, redundant mechanisms and procedures must be in place to ensure passenger and worker safety.

Another example of an electrical device that will interrupt power to the driving-machine motor is an electrical fuse or circuit breaker, which may range from a large circuit breaker at disconnect in the machine room to a fractional-amp fusible link in a small glass cylinder mounted on a circuit board. These protective over-current devices are intended to interrupt the electrical current and stop a motor or other device before it is damaged or causes an electrical fire.

There are many other elevator functions that can fail or enter a potential fault state, and in most instances the motion controller, shown in Figure 5-6, will respond by removing power from the driving-machine motor and brake, in which case the car will stop and remain stationary until the fault is cleared. Sometimes clearing the fault will automatically (without human intervention) restart the motor so that the car resumes its trip to the next scheduled stop. An example in older elevators is when the hoistway door is not latched securely due to a mechanical fault in the door hardware. Pressing the door firmly from the outside so that it clicks will often cause the motor to restart. Other faults require that the motion controller be reset. This can be done by pressing a very small (usually red) button on the motion controller's main circuit board. The same thing can be accomplished by powering down the system at the disconnect, then switching it back on.

Modern motion controllers are equipped with an alphanumeric readout, which may give an error code. These error codes are proprietary and most manufacturers do not make them available on the internet. An exception is some older models. Generally if you are dealing with a specific make and model, documentation accompanying the elevator at the time of installation should list the many error codes with recommended corrective action.

FIGURE 5-6 To reset the motion controller, press the small red button on the printed circuit board on the left. (*Judith Howcroft*)

This is what error codes look like in an elevator service manual:

Error Code: 400

Probable cause and corrective action: Motor overload. Check for dragging elevator brake or weakened motor field. A motor overload fault does not automatically shut down the drive, but it is annunciated via the ALARM relay output K2.

Error Code: 405

Probable cause and corrective action: the Safety circuit is not closed. The Drive has detected that the Safety Circuit wired between TB3-1 and TB3-6 on the Power Supply A4 was not closed for 100 Ms before a Drive Run command was given, or that it opened unexpectedly while the loop contactor was closed. Check for inter-mittent connections in the Safety Chain.

It is to be emphasized that these error codes and probable causes and corrective actions are not universally applicable among the many elevator makes and models. Elevator technicians need to obtain the service manual pertaining to the exact make and model. Then they can be assured that error codes are valid.

Most elevator service manuals are available only to authorized users and error code information is a closely-held secret. It is a matter of debate whether the purpose of restricting access to this material is to protect the public from mistakes made by untrained workers, or to restrict its availability to very high-priced proprietary service providers. An exception is the Magnetek 2004 Motion Controller Service Manual, which can be downloaded as a PDF at http://www.elevatorcontrols.com/pdfs/manu-als/DRIVES/cs0407.pdf .

It is recommended that you download and carefully study this manual to obtain an overview of a representative elevator motion controller, bearing in mind that it is not directly applicable to other makes and models.

An elevator service manual ordinarily contains a complete multi-page electrical schematic of the entire elevator system. Throughout this system as shown in the schematic are numerous sensors, such as devices that report the speed and direction of the car motion, temperature of the motor, and status of the car and hoistway doors, etc. The sensors are wired to the motion controller, and if a parameter is interpreted as abnormal, either electrical power will be removed from the motor and brake causing the car to stop in place or, if the anomaly is not hazardous but requires maintenance, an alarm may be activated while the elevator is allowed to continue operating. Many of the sensors are connected to the motion controller via two conductors, usually run in electrical metallic tubing (EMT).

A fault signal communicated to the motion controller can be due to a real elevator malfunction, often a potentially hazardous situation. Other scenarios are malfunction of the sensor, degraded electrical termination caused by improper torque applied during installation or hardware failure, damage to the wire (short, open, high-impedance caused by abrasion) or, rarely, component failure within the motion controller.

When an elevator motor stops, it is the task of the repair technician to find the cause and make a repair such that the problem does not recur and *without introducing another hazard.*

The motion controller is the elevator's brain. Before the mid-twentieth century, it consisted of electromechanical relays, wired with soldered terminations. Currently, if you open the front panel cabinet door, you'll see printed circuit boards with lots of semiconductors including integrated circuits. These devices and supporting circuitry are overwhelmingly proprietary, not generic. What they have in common is performance requirements, but how these are confronted varies widely among the many manufacturers.

Typically, the many sensors such as those that report and verify car motion and door status are wired to the motion controller, which on the basis of these inputs as well as in response to push-button controls in the car and at landings, decides car motion (speed, direction, start/stop), door operation, user notification, and the like.

If an input configuration leads the controller to conclude that there is a potentially hazardous condition, the usual response is to remove power from the motor and brake, causing the car to stop.

Lobby attendants generally call facility electricians to investigate, and often the problem can be solved without calling in outside elevator specialists.

In an ideal world, the elevator would not be restarted until the failure was completely diagnosed and repaired. Realistically, the reset button is usually pressed and operation resumed as if nothing had happened. The problem with this approach is that it is a shot in the dark, since a recurrence can mean more extensive damage. This precise judgment is made many times per day all over the world.

The motion controller can provide for selective collective operation, or group automatic operation. In selective collective operation, the elevator control system retains all calls. It answers those intending to travel in one direction first, then reverses the car direction and picks up the others. Finally, the car returns to the designated landing.

Group automatic operation is a more complex protocol. It is used in a group installation where individual cars respond in the most efficient manner as directed by the motion controller.

The elevator control system is comprised of inputs, outputs, and one or more controllers. Sensors, one class of inputs, are often the primary focus in elevator diagnosis and repair. The car position is ascertained in the motion controller by means of a magnetic photo-electric sensor located on the car. The car position is determined by counting the number of pulses generated as the sensor passes holes in one of the guide rails.

Infrared sending units and sensors located on opposite door jambs detect passengers entering or leaving the car and prevent doors from prematurely closing on them.

A weight sensor is attached to the car. It notifies the motion controller when an overload condition exists.

An encoder placed at the drive sheave is read by the primary velocity transducer, which is wired to the motion controller.

Hall buttons, floor request buttons, the open door button, stop button, ring bell button, mode switch, emergency stop button, and emergency button may all be considered sensors. This category also includes key switches such as Fireman's Service, inspector's switch, car preference switch, and independent service mode switch. The status of these switches may be indicated in the motion controller alphanumeric readout.

The car door opening device is located on the top of the car. In older elevators, it may be driven by a fractional horsepower motor equipped with pulleys and V-belt. Now, expect to see a linear actuator.

Besides enabling the elevator system to operate efficiently and facilitating diagnosis and repair, the motion controller is designed to support numerous safety mechanisms. Like the familiar circuit breaker in an electrical service or load center, many of them resemble the trigger of a firearm. What this means is that they are more easily actuated than reset. In the event of a hazardous condition, we want them to quickly and infallibly perform the intended action. The consequence of this design intent is that they may at times inadvertently trip, an inconvenience but a sign that they are poised to act instantly.

Accordingly, when the elevator stops running for no immediately apparent reason, repair technicians usually end up looking at sensors and associated wiring.

Because of variations among the many manufacturers, it is invariably necessary to look at the documentation. Unfortunately, unless extensively annotated at the time of installation, schematic drawings may not indicate unique details pertaining to the installation, such as conductor routing from motion controller to sensors. This wiring may be run in Type EMT or MC cable through the hoistway, or it may be concealed behind wall and ceiling finish. In Chapter 6, Electrical Troubleshooting Tools and Instruments, we will discuss some basic and inexpensive instrumentation that is valuable in detecting and locating faults in sensor circuitry. In Chapter 8, Elevator System Overview, we will discuss electrical codes as they pertain to knowing where to find compliant wiring in commercial and industrial settings.

STUDy QUESTIONS

1. ASTM A17 requires that sheaves and drums be made of:
   A. brass
   B. galvanized steel
   C. cast iron or steel
   D. cast iron only

2. Stopping switches at the driving machine:
   A. may be driven by chains
   B. may be driven by ropes
   C. may be driven by belts
   D. are not to be driven by chains, ropes, or belts

3. For passenger elevators, the brake is to be designed to hold the car at rest with an additional load up to _____ percent in excess of the rated load.
   A. 10
   B. 25
   C. 30
   D. 50

4. Normal stopping devices are to be located:
   A. in the car
   B. in the hoistway
   C. in the machine room
   D. in any of the above

5. Broken rope switches are to be located in:
   A. the machine room
   B. the car
   C. the hoistway
   D. any of the above

6. Hoistway door interlocks are to be provided for:
   A. traction elevators
   B. hydraulic elevators
   C. freight elevators
   D. all elevators

7. A disconnect switch or circuit breaker is to be connected in the power supply line to all:
   A. elevator motor or motor generator sets and controllers
   B. elevator hoistways
   C. elevator cars
   D. elevator pits

8. In a hydraulic elevator, the pump relief valve is to have sufficient size to pass the maximum capacity of the pump without raising the pressure more than _____ percent above the working pressure.
   A. 5
   B. 10
   C. 25
   D. 50

9. In a hydraulic elevator, a pump relief valve is not required for centrifugal pumps driven by induction motors provided the shutoff or maximum pressure which the pump can develop is not more than _____ percent of the working pressure at the pump.
   A. 125
   B. 135
   C. 150
   D. 175

10. In hydraulic elevators, tanks are to be provided with a pressure gage that will indicate pressure to not less than _____ percent of the pressure setting of the relief valve.
   A. 100
   B. 125
   C. 150
   D. 175

For answers, go to Appendix A.

CHAPTER 6

ELECTRICAL TROUBLESHOOTING TOOLS AND INSTRUMENTS

Electrical diagnostic instruments vary in cost from a simple plug-in socket with an appliance bulb that can be put together for a couple dollars to an advanced spectrum analyzer priced over $20,000. (We will demonstrate a spectrum analyzer in this chapter that can be configured with an owner-supplied laptop PC for a small fraction of that amount.)

Test Lights

Electricians use test lights, as shown in Figure 6-1, to check for presence or absence of voltage in equipment and in branch-circuit receptacles. They are great for working in large rooms and long hallways if you want to keep watch from a distance on the status of a wiring run at different points. One problem in using a multimeter for this task is that you never know for sure when the probes are making electrical contact in the slots. It's good to have two or three of these testers on the job. One advantage of the small appliance bulb is that it is less breakable than a standard bulb.

The bulb and socket equipped with wire leads stripped at the ends is good for checking for presence or absence of voltage at terminals or in tight places. After finding a presumed de-energized circuit, to be safe, the tester should be verified on a known live circuit. Another variation is the socket with leads terminating in alligator clips.

Another similar device is the neon-bulb tester. It comes with insulated leads. (It's rated for 600 volts, but that is too close for comfort for me.) The tester is excellent for identifying hot versus neutral conductors. If either lead contacts an energized object while the other lead floats, the bulb will glow weakly. For this reason, it is good for checking a chassis or enclosure for internal ground faults.

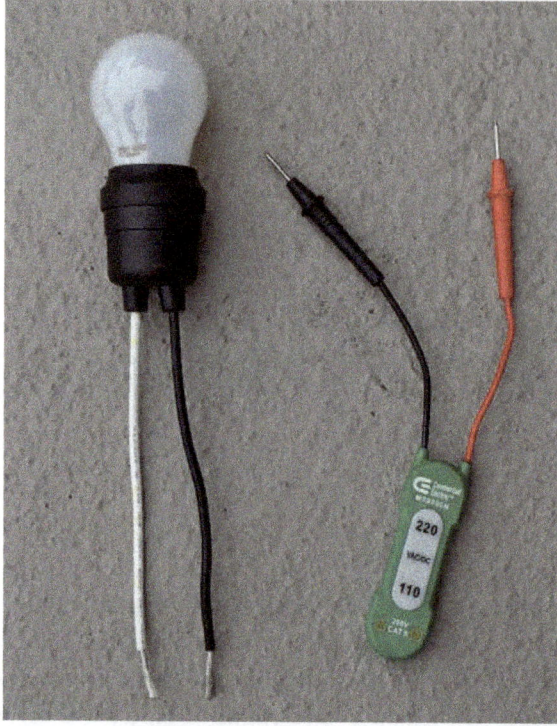

FIGURE 6-1 Appliance bulb in socket equipped with leads (left) and neon-bulb tester (right). (*Judith Howcroft*)

A circuit analyzer, shown in Figure 6-2, will check and display by means of three LEDs the status of wiring supplying power to a receptacle. This little plug-in instrument permits the user, by reading the LED pattern, to identify various wiring faults, including no power to the receptacle, missing equipment ground, hot and neutral reversed, or wiring is correct. The NEC, in Article 620 Elevators, Dumbwaiters,

FIGURE 6-2 The circuit analyzer is designed to check old work and verify new branch circuits. (*Judith Howcroft*)

Escalators, m oving Walks, Lifts and Chairlifts, states that at least one 125-volt, 15- or 20-amp duplex receptacle is to be provided in each machine room or control room and machinery space or control space. An identical requirement applies to the pit. In performing routine maintenance, elevator technicians should verify that these receptacles are in place. They should be on dedicated circuits, and they should be checked with the circuit analyzer to make sure that reliable power is in place in the event of an elevator outage.

Multimeters

The electrical testers described above are convenient for quick preliminary troubleshooting and checking for the presence or absence of branch-circuit voltage, but for serious diagnostic work a high-quality multimeter is essential. Big-box outlets and hardware stores carry small digital multimeters that sell for around $10. This type of meter is OK for backup or a spare, but it is neither durable nor reliable and is not suitable for serious diagnostic work.

Fluke offers a wide range of high-quality multimeters. They are available online from multiple reputable dealers. For the elevator technician, an excellent model is the Fluke 87V industrial True Rm S m ultimeter with temperature measuring capability, shown in Figure 6-3. It sells for under $400.

FIGURE 6-3   Fluke 87V multimeter. (*Fluke*)

This multimeter is similar to Fluke's classic model 87 meter, but it has advanced features applicable to VFD AC motors, as seen in most modern elevator installations. Also, the 87V has improved measurement functions, troubleshooting capabilities, resolution, and accuracy. All inputs are protected to Category III, 1000 volts, and Category IV, 600 volts.

WHAT'S ALL THIS ABOUT WIRING CATEGORIES?

Electronic test equipment is always marked CAT I through CAT IV with voltage ratings. These categories correspond to locations in which higher amounts of available current exist. The higher the CAT number, the greater the electrical hazard.

CAT IV consists of outdoor locations, notably three-phase utility terminations, aerial lines to detached buildings, service-entrance installations, and underground lines to connected loads.

CAT III consists of indoor three-phase distribution, large equipment such as three-phase motors in fixed locations, industrial bus and feeder installations, large lighting loads, and load centers or appliance outlets located near entrance panels.

CAT II consists of single-phase branch circuits and loads, outlets more than 30 feet from CAT III sources, and outlets more than 60 feet from CAT IV sources.

CAT I consists of protected electronic gear including high-voltage, high-impedance office equipment such as paper copiers.

Test equipment is rated for the maximum voltage that can be safely contacted in each location. Typically, there are multiple ratings such as 1000 V CAT III, 600 V CAT IV. These CAT ratings do not correspond to OSHA electrical work zone boundaries within which different personal protective equipment is required.

They can withstand impulses in excess of 8000 Volts and there are reduced risks related to surges and spikes. Very accurate voltage and frequency measurements can be taken on 480-Volt VFDs, and in AC only mode, ripple can be evaluated on the DC bus.

Where this multimeter differs from other high-performance meters on the market is that it has a yellow button on the front panel that, by pressing it, permits the user to temporarily reduce the instrument's bandwidth. This is necessary for accurately measuring a VFD output. That output consists of quasi-square wave pulses of varying duty cycles that generate strong high-frequency harmonics, which you do not want to include in measuring the electrical power at the VFD output/motor input.

The 87V multimeter also has a built-in thermometer that allows the user to take temperature readings without needing a separate instrument. An optional current clamp permits the user to take high-amp readings without breaking open the circuit.

This multimeter also performs resistance, continuity, and diode tests. It has a lifetime warrantee.

Non-Contact Meters

A variation on the familiar multimeter is the Fluke T6–600 non-contact electrical tester, shown in Figure 6-4.

FIGURE 6-4 Fluke T6–600 non-contact electrical tester. (*Fluke*)

To measure AC volts and amps, slide the insulated hot wire (black) into the fork so that it bottoms out. Volt and amp values are shown simultaneously in the digital readout. Press the yellow button to read frequency in Hz.

The user must supply a capacitive path to ground, at which point the display turns green, indicating a valid measurement. The path to ground is enhanced by either touching the black lead to a conductive grounded object such as a chassis enclosure, or touching the field sense on the front panel with your finger.

AC and DC can be tested up to 1000 volts. Resolution is one volt.

Use of this non-contact meter is far less hazardous than a conventional multimeter with respect to shock and arc flash because measurements are taken on insulated conductors making use of capacitive coupling only. Safety ratings are 1000 volts Cat III and 600 volts Cat IV.

Another frequently used electrical test instrument is the clamp-on ammeter, such as the Fluke 323, seen in Figure 6-5.

FIGURE 6-5   Fluke 323 clamp-on ammeter. (*Fluke*)

Priced around $130, this instrument makes quick, safe current measurements up to 400 amp. The Cat III rating is 600 volts; Cat IV rating is 300 volts. High-current measurements can be quickly and safely taken without breaking or affecting the operation of a working circuit.

The jaws of the instrument are opened and a current-carrying conductor is inserted. The wire does not have to be precisely centered or perpendicular to the opening. The clamp-on ammeter measures the magnetic field surrounding the energized conductor connected to a load and the digital readout displays the amount of current flowing through it. For small amounts of current outside the meter's range, the user can coil the wire and arrange it so that it passes twice (or more) through the jaws. That multiplies the current reading, so the user has to divide it accordingly.

For an induction motor, current readings are sometimes more revealing than voltage readings, and these same readings tell a lot about the power supply as well.

Power Quality Measurements

Virtually all elevator motors run on three-phase power, which has multiple advantages including higher efficiency and ease of wiring in contrast to single-phase. However, deficient power quality in a three-phase supply can set the stage for accelerated wear, if not immediate failure, in an induction motor. Preventive maintenance and failure analysis invariably involve an examination of the three-phase power, measured at various points between the facility electrical service and the VFD power input.

The first instrument generally deployed is the ubiquitous multimeter, which is suitable for measuring the all-important amplitude parameters. These metrics, while essential in determining whether the voltages in all three legs, with and without connected load, are substantially equal, do not constitute the complete power-quality picture. The oscilloscope, spectrum analyzer, energy logger, and power quality analyzer are all useful and have their place in the elevator machine room.

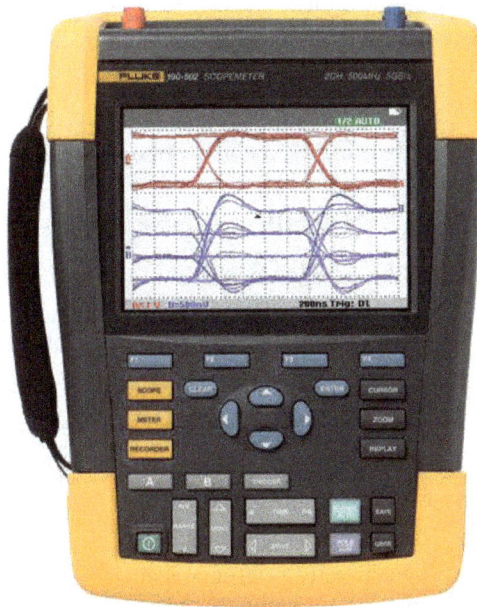

FIGURE 6-6   Four isolated inputs enable the Fluke 435 Series II Power Quality and Energy Analyzer to safely display three-phase waveforms plus neutral simultaneously. (*Fluke*)

The Fluke 435 Series II Three-Phase Power Quality and Energy Analyzer (Figure 6-6), selling on Amazon.com for just under $6,400, measures and displays numerous details pertaining to the connected three-phase power:

- Phase voltages are displayed and should be close to the utility nominal value and to the specified requirements for the installed VFD. The voltage waveforms must be a sine wave that is smooth and free of distortion. Scope Waveform displays the shapes of the three waveforms. Dips and Swells records sudden voltage changes. Transients m ode captures voltage anomalies.
- Phase currents: Volts/Amps/Hertz and Dips and Swells check current and voltage relations. Inrush current records sudden current inrushes such as motor inrush.
- Crest factor: If it is 1.8 or higher, there is waveform distortion.

■ Harmonics mode checks for voltage and current harmonics and total harmonic distortion (THD) per phase. Trend records harmonics over time.

■ Flicker is used to check long- and short-term voltage flicker and similar data in each phase. Trend records this data over time.

■ Dips and Swells records sudden voltage changes down to one-half cycle.

■ Frequency rarely differs from the utility nominal value. Volts/Amps/Hertz displays frequency. Use Trend m ode to search for frequency anomalies.

■ Unbalance reveals phase voltage differences. They should be less than one percent of the average value of the three phases. Current difference in an individual phase must be less than ten percent.

■ Energy Loss Calculator shows location and amount of energy losses.

■ Power Inverter Efficiency displays the efficiency and energy output of connected inverters.

■ m ains Signaling measures superimposed data present in the electrical supply.

■ Power Wave is a high-resolution eight-channel waveform recorder.

Oscilloscopes

Other than the multimeter, the oscilloscope is the most widely used diagnostic tool in electronics work, and this is particularly true in elevator fault diagnosis. The instrument is suitable for checking three-phase power quality at the VFD input and ripple on the DC bus. But you have to remember, as emphasized elsewhere in this book, that a conventional grounded bench-type oscilloscope (unless equipped with differential probes) cannot be used to display voltages where both sides of the circuit are referenced to but float above ground potential. In that hookup, the instant the probe ground reference lead contacts a floating voltage, hazardous current rushes through the circuit under investigation, ground reference lead, oscilloscope, and facility equipment grounding conductor, then back to the utility neutral at the entrance panel.

m ost elevator technicians, rather than using high-priced differential probes in conjunction with a grounded bench-type oscilloscope, display three-phase supply and internal VFD voltages by means of a hand-held, battery-powered oscilloscope, shown in Figure 6-7. This instrument has the advantage that its inputs are isolated from ground potential and, in most models, are insulated from one another. You can safely measure Y- and delta-configured three-phase circuits and internal VFD floating voltages such as on the DC bus without fault-current hazard.

Priced just over $5,600, this is the closest a hand-held, battery-powered instrument comes to a high-end, bench-type oscilloscope. Of course with fully-insulated inputs rated up to 1000 volts, the Scopem eter is safe for displaying the various ground-referenced floating voltages found in a 480 VAC three-phase VFD. (A more expensive four-channel model is also available.)

This Fluke test instrument differs from other hand-held oscilloscopes in that it is also a full-featured multimeter, with separate probes. Volt, amp, ohm, and Hertz values are displayed in a large, very clear readout. For most diagnostic work, this is

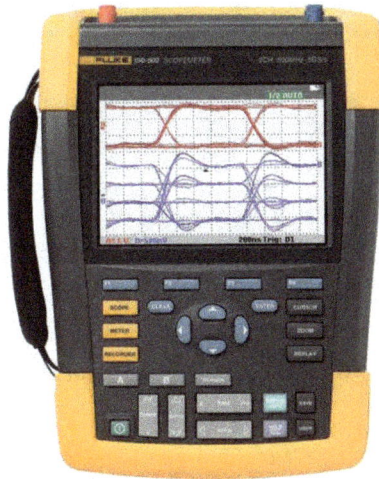

FIGURE 6-7 Fluke 190-502 ScopeMeter is a two-channel, 500-MHz instrument. (*Fluke*)

the only instrument other than a thermal imager that the elevator technician will need to bring into the machine room. On the factory floor, the Scopem eter's rugged dust- and moisture-resistant enclosure will withstand reasonable impacts and harsh surroundings, and it is suitable for extensive outdoor work as well.

The 5GS/s real-time sampling rate and 500 m Hz bandwidth, which can be temporarily limited to view noisy signals, permits the user to test a wide range of micro-electronics. These same specifications mean that the instrument is suitable for viewing the pulse-width modulated control signal that is generated within the motion controller and is conveyed to the data inputs of the IGBTs in the VFD inverter section. These signals, then, can be compared individually to each of the semiconductor outputs, and to the individual three-phase power inputs at the motor terminals.

The Scopem eter is also useful for displaying sensor inputs at the motion controller, in order to detect noise and electromagnetic interference that might not be picked up by a multimeter.

A full-featured bench-type oscilloscope, such as the Tektronix m DO3000 Series, shown in Figure 6-8, is a highly capable instrument with seemingly endless display and analysis features.

It can be brought into the elevator machine room under certain conditions:

- It is equipped with differential probes so that ground-referenced floating voltages can be safely measured and displayed. The elevator technician must be fully knowledgeable regarding where these voltages may be encountered, why they are hazardous in relation to a grounded instrument, and proper use of the differential probes to mitigate the hazard.

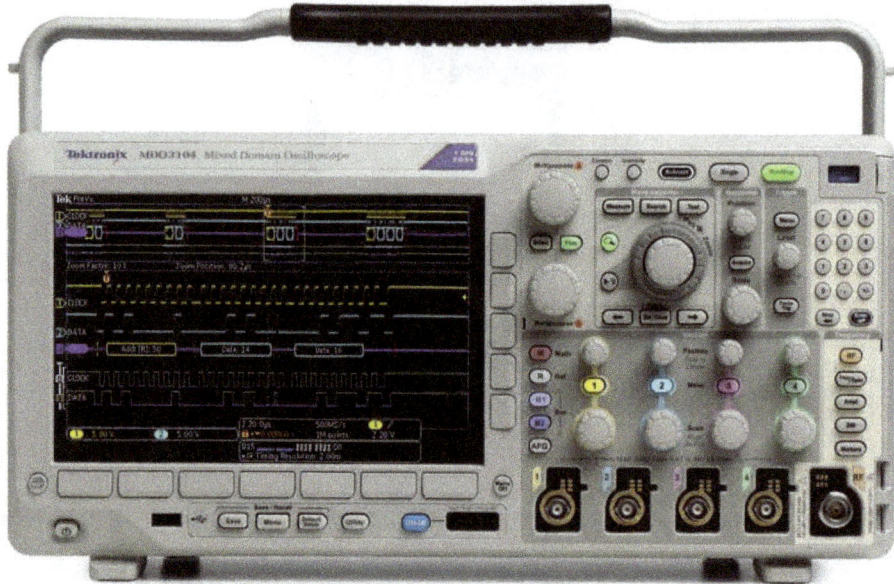

FIGURE 6-8   Tektronix MDO3104 Oscilloscope. (*Tektronix*)

■ A clean, well-lighted table or suitable work surface is located adjacent to the motion controller. It should be large enough to accommodate the oscilloscope with sufficient space to spread out schematics and related documentation. A grounded receptacle is essential for powering the oscilloscope, other instruments, and tools. (Note: Any grinding, cutting, drilling, or soldering operations should be performed in another room so that conductive particles cannot contact circuit boards and terminations inside the motion controller enclosure, not to mention the oscilloscope.)

The m DO3000 has digital as well as analog channel inputs. Any digital signal should first be examined as an analog signal to detect the presence of noise or distortion. Problems of this sort often have to do with cabling or terminations.

The various ways of displaying data in the digital channels permit the user to analyze the signals. Digital channels store a high or low state for each sample. Logic high levels are displayed in green and logic low levels are displayed in blue. When a single transition occurs during the time represented by one pixel column, the transition (edge) is displayed in gray. When multiple transitions occur during the time represented by one pixel column, the transition is displayed in white. When the display shows a white edge, indicating multiple transitions, you may be able to zoom in to see the individual edges.

The m DO3000 oscilloscope can display up to 16 digital channels simultaneously. You may wonder, considering the fact that going from two to four analog channels increases the cost of an oscilloscope dramatically, how it is possible to display so many digital channels. The answer is that these signals are already digitized, so that the costly clock, ADC, and sampling hardware is not needed for each channel. The digital signals are presented as-is, directly to the processor.

The oscilloscope is able to display the following digital bus protocols:

- Parallel (mostly obsolete)
- I²C
- SPI
- RS-232
- CAN (frequently used in modern elevator systems)
- LIN
- Flex Ray
- Audio
- USB
- m IL-STD-1553

We will discuss digital servicing techniques as applicable to elevator systems in Chapter 9.

The arbitrary function generator (AFG), shown in Figure 6-9, is a very basic supplement to the oscilloscope. Like most digital storage oscilloscopes, the model shown has an internal AFG, which is useful, particularly in order to provide a sample waveform so that the many capabilities and advanced features of the oscilloscope can be demonstrated. A well-equipped shop, however, should have a stand-alone, bench-type AFG so that signals can be generated, modified, and injected at various points into equipment under investigation. Then, the oscilloscope can proceed, stage by stage,

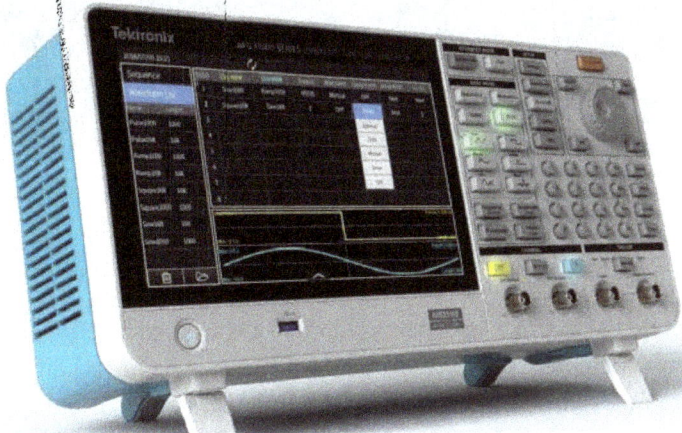

FIGURE 6-9   Tektronix AFG31000 Arbitrary Function Generator. (*Tektronix*)

circuit by circuit, component by component, until the deficiency is located. This is the essence of electronic troubleshooting, and it can be customized to locate faults in an elevator system, within and outside the motion controller.

The new Tektronix AFG 31000 Arbitrary Function Generator is unique in that the Home screen, accessed by pressing the Home button on the front panel, permits the user to choose a Basic or Advanced mode. The Basic mode resembles a traditional AFG, with a library of internal waveforms and menus that permit the user to adjust parameters such as frequency and duty cycle. The Advanced mode makes available enhanced features that facilitate the many tasks that today's electronics engineers tackle in the laboratory, shop, or in the field.

The very large touch screen has prominent tabs for choosing Basic or Advanced, in addition to smaller tabs that bring up ArbBuilder, Utility, and Help. Users will be eager to try out the new ArbBuilder with enhanced features, but we will start with some fundamental functions (Figure 6-10).

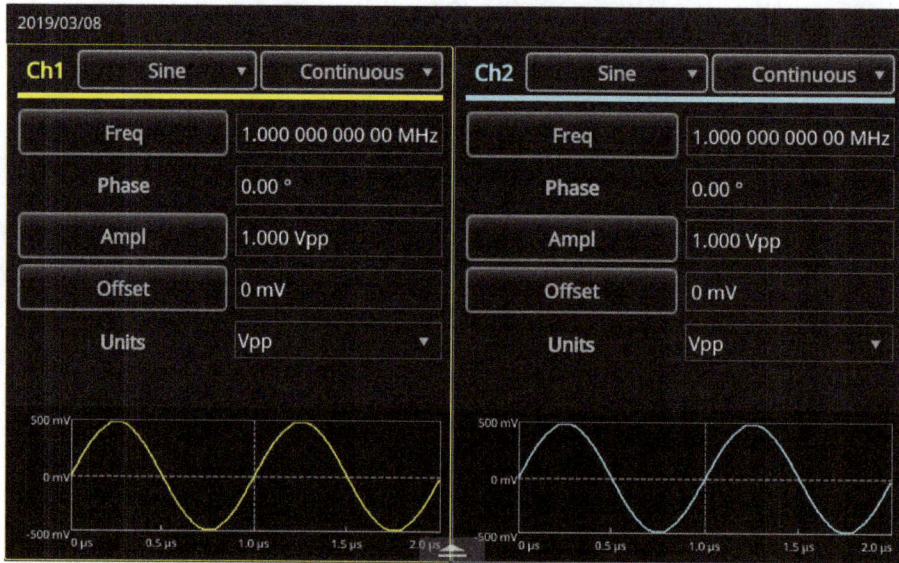

FIGURE 6-10 The Basic functions screen permits the user to select a waveform and set parameters.

When pressing Basic, in a two-channel model, we see in split-screen format a drop-down menu containing the internal library of waveforms permanently stored in the instrument's memory. A second drop-down menu specifies whether the selected waveform is to be displayed in continuous, modulated, swept, or burst mode. To see the functions displayed with modifications, in the AFG front panel channel section, as shown in Figure 6-11, insert two BNC cable connectors into the Channel One and Channel Two output ports and run them to a digital storage oscilloscope such as the Tektronix m DO3000, connecting two analog channel input ports.

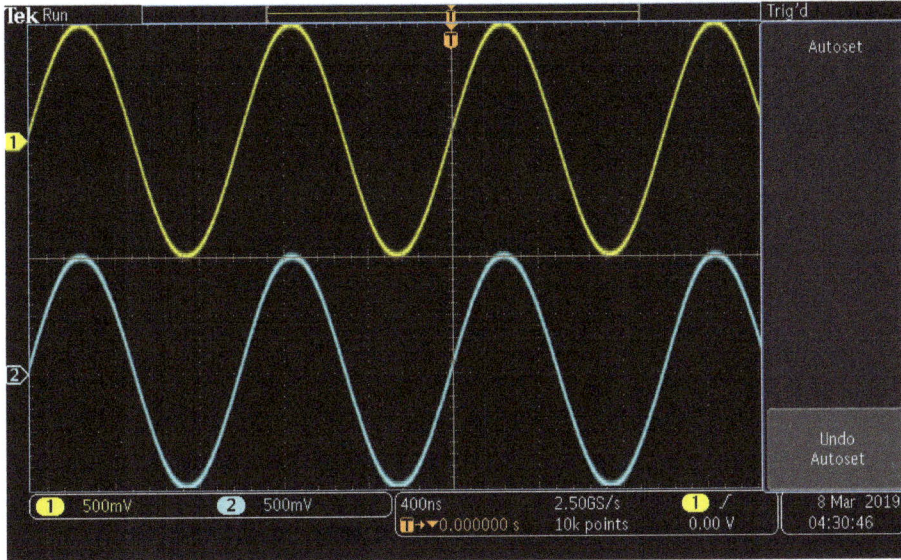

FIGURE 6-11   When the two AFG channels are on and the two oscilloscope channels
are on with BNC cables connecting the two instruments,
the two waveforms are displayed in the oscilloscope.

Accessing the two internal waveforms shown in the two AFG drop-down menus, the user can select twelve alternate functions plus Arbitrary. The same or different waveforms can be selected independently and sent to the oscilloscope. There, they are redigitized and can be stored, analyzed, modified, or emailed to a colleague or publisher.

For example, turn both channels on in both instruments. With ramp in Channel One and square wave in Channel Two displayed, press the m ath button on the oscilloscope front panel. The horizontal m ath menu appears with default Dual Waveform active in the horizontal m ath menu. Pressing the associated soft key, we see that sources are Channel One and Channel Two and that the operator is set at Add. The sum of the two waveforms is shown in red, the dedicated m ath color (Figure 6-12).

Now turn off Channel Two in both instruments and turn off m ath in the oscilloscope.

Notice in the AFG Channel Two basic display that the square wave is still shown, and the various waveforms can be selected and modified. That is because the on-off button is applicable only to the channel output.

In the AFG, turn the Channel One waveform back to Sine. The oscilloscope display responds instantly. Now we'll see how waveform metrics correspond in the two instruments.

In the AFG, the middle part of the Basic mode screen shows waveform parameters. They are Frequency, Phase, Amplitude, Offset, and Units. If you touch Frequency, the displayed value shifts to the reciprocal, Period. The Frequency/Period of the signal at the AFG output does not change for now. Similarly, touching Amplitude, Offset, and Units changes the way the waveforms are measured, but not the actual signals.

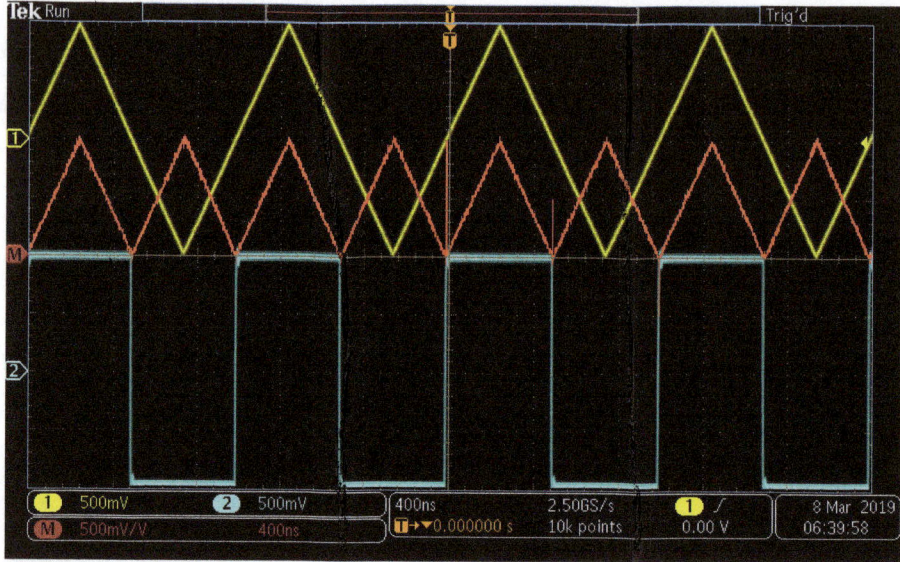

FIGURE 6-12 Operators can be toggled, using the soft key.
Here Divide is shown in the oscilloscope.

Signal parameter can, however, be changed using buttons in the Basic m ode front panel controls in conjunction with the front panel or on-screen number pad, whichever is more convenient for you. For example, to change the frequency of the sine wave to 2 m Hz, press the Frequency/Period button in the front panel Basic m ode section. In the number pad or using the touch screen, enter 2. The frequency changes to 2.000 m Hz. Similarly, Amplitude can be changed.

Pressing Default in the Setup Section, the frequency reverts to 1 m Hz and the amplitude is 1.000 Volt peak-to-peak.

In the oscilloscope Wave Inspector section on the front panel, press m easure. In the horizontal m easure menu, press the soft key associated with DVm and use m ultipurpose Knob A to scroll down to Frequency. As expected, the frequency measures a stable 1.0000 m Hz, exactly the same as the AFG, but carried to fewer decimal places, as shown in Figure 6-13.

Don't expect the amplitudes to correspond at this frequency. There is too much loss in the BNC cables and the oscilloscope DVm reports Over Bandwidth (Figure 6-14).

The Tektronix AFG 31000 series offers four modes for viewing the internal waveforms. They are Continuous, m odulation, Sweep, and Burst, shown in Figure 6-15. All waveforms do not work in all modes. Figure 6-15 is a summary.

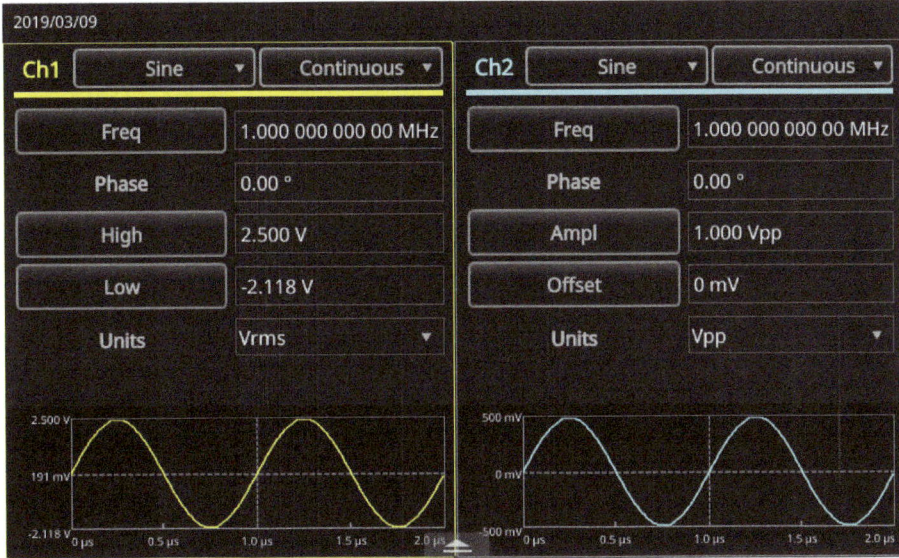

FIGURE 6-13   One MHz sine wave from the AFG.

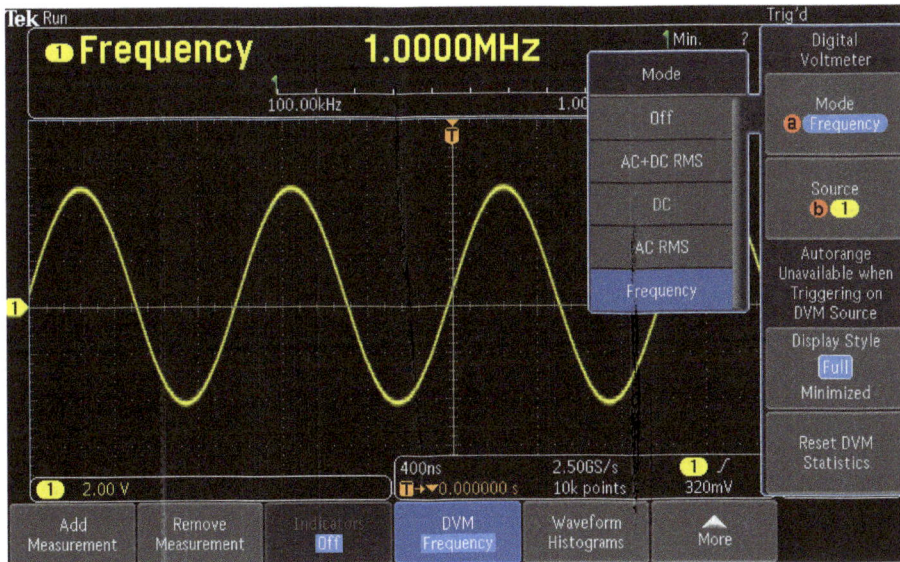

FIGURE 6-14   One MHz sine wave displayed in the oscilloscope with frequency measured.

Run mode	Sine, Square, Ramp, Arb, Sin(x)/x, Gaussian, Lorentz, Exponential rise, Exponential decay, Haversine	Pulse	Noise, DC
Continuous	•	•	•
Modulation			
AM	•		
FM	•		
PM	•		
FSK	•		
PWM		•	
Sweep	•		
Burst	•	•	

FIGURE 6-15 Noise and DC work only in Continuous Mode. No waveforms can be accurately displayed in an external oscilloscope in modulation mode, because two waveforms would be required.

m odulation can consist of amplitude modulation (Am), frequency modulation (Fm), as shown in Figure 6-16, phase modulation (Pm), and frequency shift keying modulation (FSK).

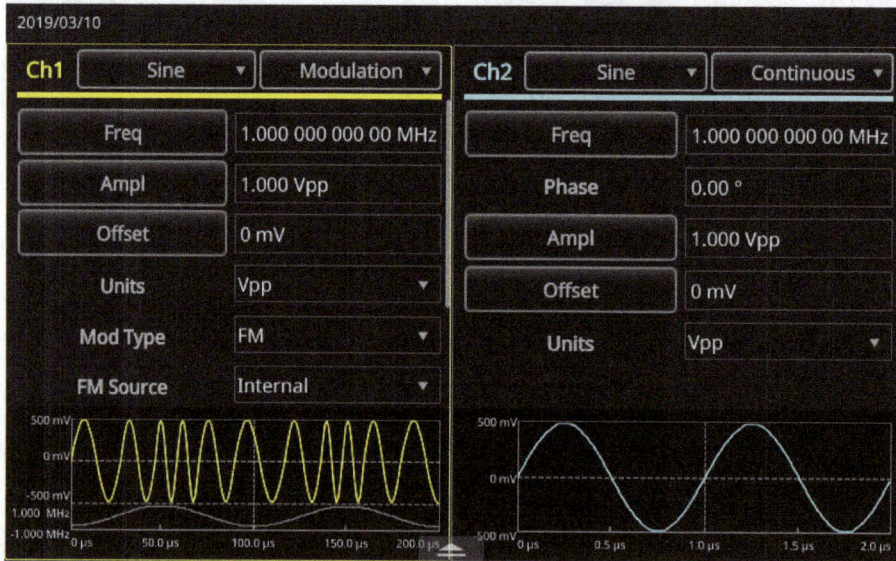

FIGURE 6-16 FM modulation.

Besides modulation, signals may be swept or in bursts. The swept signal, as shown in Figure 6-17, varies linearly or logarithmically. Settings include Start Frequency, Stop Frequency, Amplitude, Offset, Units, Sweep Time, Return Time, Center Frequency, Frequency Span, and Hold Time.

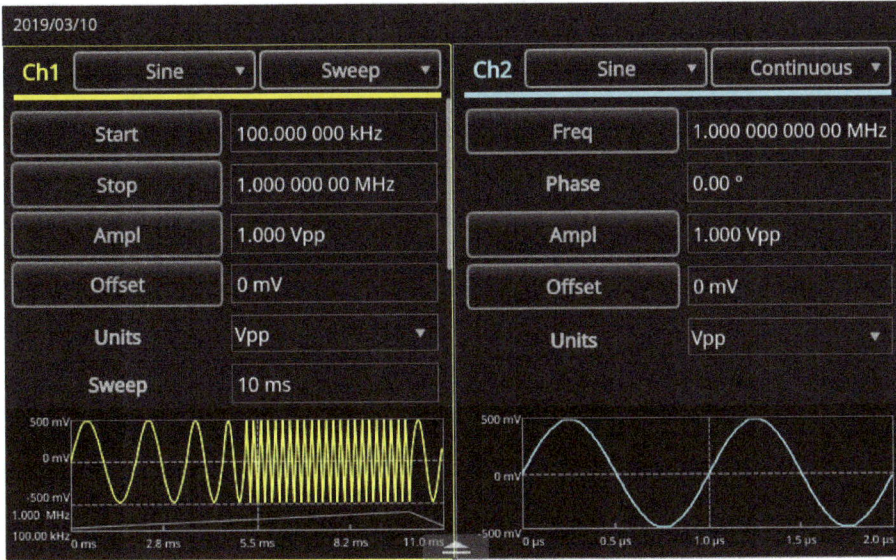

FIGURE 6-17   Swept sine wave.

In Basic m ode, the AFG can also output a Burst Signal. The instrument provides these trigger sources to enable the burst mode: internal or external trigger signal, manual trigger, and remote command.

A screen image can be saved to a USB flash drive. Follow these steps:

1. Insert a USB flash drive into the front panel USB slot.

2. Set the display to show the screen you want to save.

3. Simultaneously press the two arrow keys underneath the navigation control on the front panel.

4. A message appears on the screen indicating the image was saved.

The AFG 31000 in Advanced m ode, shown in Figure 6-18, can perform a great many complex operations with a relatively intuitive and user-friendly interface. The focus in this AFG Advanced m ode is that it enables users to generate long waveforms and those with complex timing. While it is true that the AFG Basic m ode, and for that matter even the internal AFG in the Tektronix m DO3000 both have arbitrary waveform generation capability, the AFG 31000 in Advanced m ode introduces enhanced features. Users can control how the waveforms in Sequencer are organized, including repeat, wait, jump, go to, or triggering with events in external trigger, manual trigger, timer, and SCPI commands.

In Advanced m ode, every point in an arbitrary waveform is output once each cycle at the pace of the specified sampling rate, free of skips or repetition. Advanced m ode is appropriate when jitter and phase noise cannot be tolerated, as in serial bus simulation, or when narrow anomalies are to be simulated in long waveforms, for example to test prototypes for such vulnerabilities. In this situation, the waveform

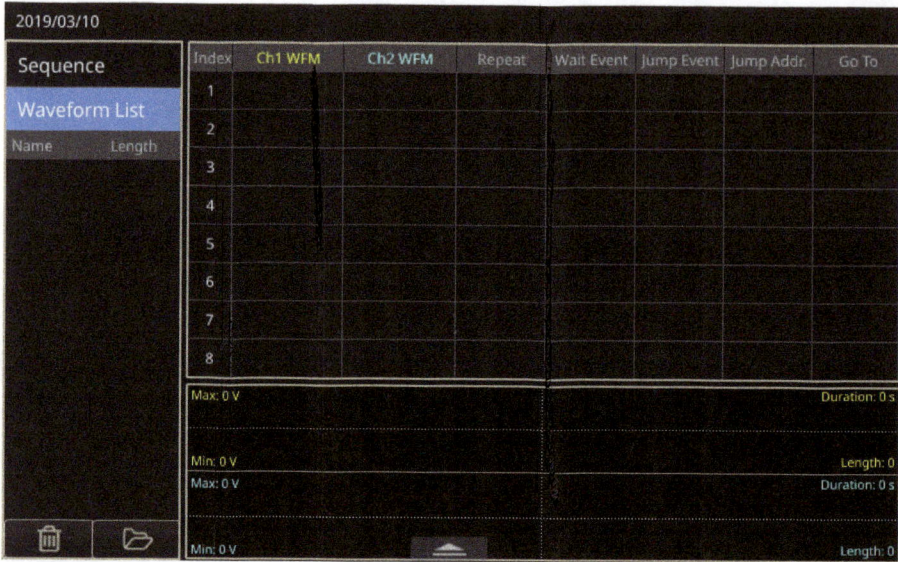

FIGURE 6-18   The AFG Advanced Mode default page. As arbitrary waveforms are generated, they can be archived in tabular form and later accessed.

details are retained, jitter and phase noise are reduced, and in each cycle the number of samples is consistent.

Advanced mode has four output modes: Sequence, Continuous, Triggered, and Gated. Sequence offers enhanced flexibility when generating waveforms characterized by complex timing. In the Continuous output mode, the AFG 31000 outputs an ongoing one-step waveform sequence. In Triggered, one cycle of a waveform is output when the AFG receives a trigger input. After one cycle, the instrument returns to its initial state until another trigger input is received. In Gated mode, the AFG outputs a one-step waveform sequence when it receives a valid gate signal, then stops upon receipt of another gate signal.

The Advanced mode outputs the waveform in accordance with the sequence table, including loop and conditional jump. The waveform is output as defined by the sequence that is selected. A variety of waveforms can be output in any specified order.

The Advanced output screen permits the user to choose from available waveforms, open a saved waveform by selecting Sequence, and create and save a new waveform.

Closely related to the oscilloscope is the spectrum analyzer, shown in Figure 6-19. The enclosures are similar, and both have prominent screens with numerous buttons and controls.

As in the oscilloscope, channel inputs are at the bottom of the front panel.

A modern digital storage oscilloscope is capable of displaying signals in time or frequency domain. In contrast, the spectrum analyzer is optimized to display signals only in the frequency domain. Despite this limitation, it has numerous advanced features that provide a great amount of information pertaining to the signal at its input.

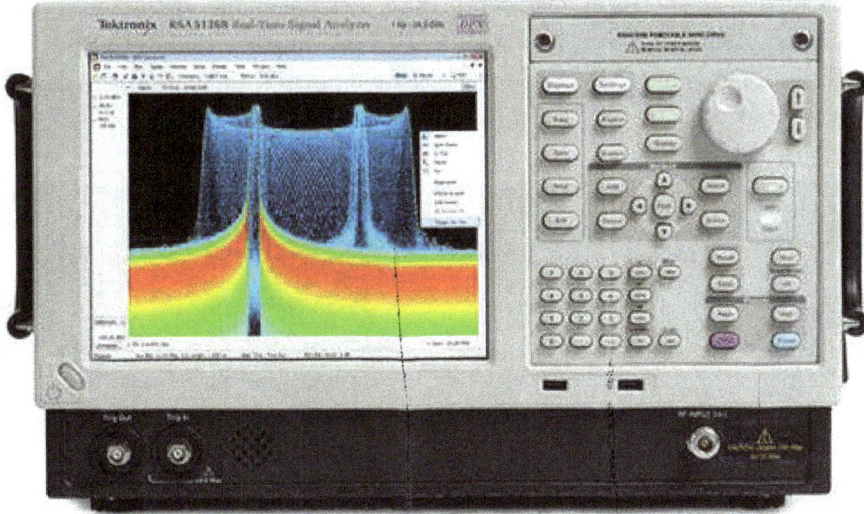

FIGURE 6-19   Tektronix RSA5000 Real-Time Spectrum Analyzer. (*Tektronix*)

In the frequency domain the spectrum analyzer surpasses the oscilloscope, often providing insights that help solve a difficult troubleshooting problem.

Both time-domain and frequency-domain displays consist of graphs of the electrical signals at the input. The time-domain display depicts amplitude in volts plotted against the vertical Y-axis and time in seconds plotted against the horizontal X-axis. The frequency-domain display also depicts amplitude, now in decibels rather than volts, against the Y-axis. Rather than time, it depicts frequency in Hertz against the X-axis. The same signal is shown, but it has a whole different look on the screen, as shown in Figure 6-20.

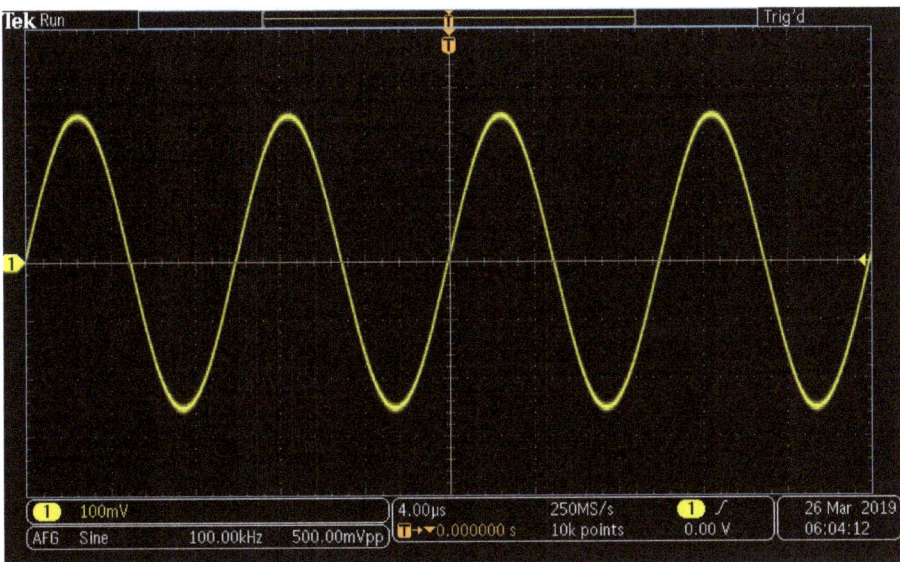

FIGURE 6-20   Sine wave in time domain.

A bench-type spectrum analyzer displays a signal at the instrument's input in the frequency domain, shown in Figure 6-21, providing for the user the dominant frequency, amplitude in terms of power of the fundamental and harmonics, distortion, noise content, bandwidth, and other properties.

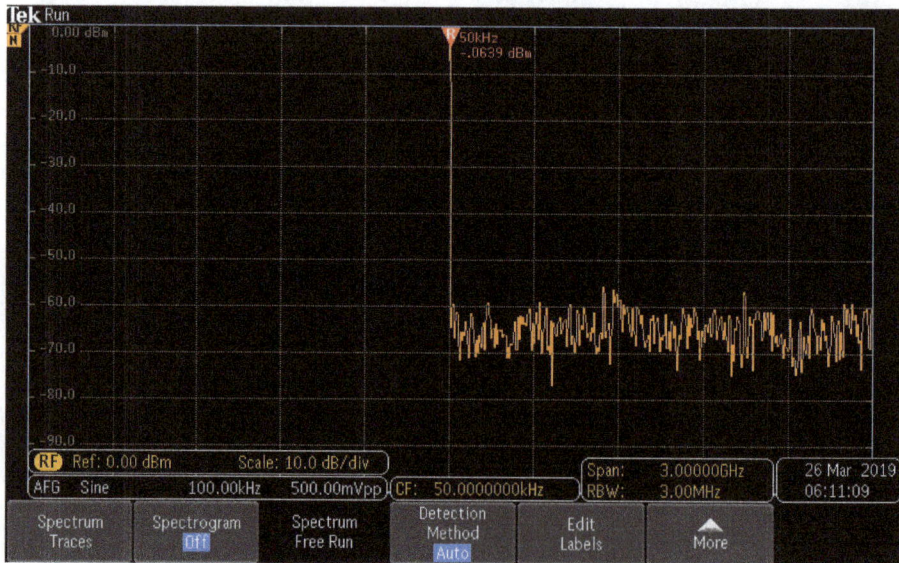

FIGURE 6.21   Sine wave in frequency domain.

At first, even an experienced oscilloscope user may be at a loss when confronted by the array of front panel controls and winding menu paths. It may be difficult to display the fundamental and even a square wave, with its near instantaneous rise and fall times, may not appear to have any harmonics.

First, you have to understand the irregular, rapidly fluctuating horizontal trace near the bottom of the screen. It is the noise floor of the instrument. It also appears in the oscilloscope when it is in FFT or real-time frequency analyzer mode.

All conductors, including components such as resistors as well as active semi-conductors, if not at absolute 0 degrees Kelvin, are subject to thermal noise due to random motion of charge carriers such as electrons and ions. It is this thermal noise that appears as phantom voltage in an auto-ranging multimeter when, in the volt mode, it is not probing an energy source.

A very small but measurable amount of voltage will appear across the open leads of any conductor or component that is not in circuit, and if the leads are shunted, a small but measurable current can be detected by appropriate instrumentation.

This noise floor is always present in the display of an auto-ranging spectrum analyzer. It is not a defect in the instrument, but a consequence of its great sensitivity. No signal or signal component with an amplitude below the noise floor can be displayed.

As previously mentioned, at first there may be a problem displaying a signal in the spectrum analyzer. The fundamental is difficult to see, because it coincides with the left edge of the display. To bring it to the center, carefully note the frequency of the square wave coming out of the AFG. It is 100 kHz, the default. Press the Freq/Span button, which brings up the Frequency and Span menu. You can see that currently the Center Frequency is 1.50 GHz, the spectrum analyzer default, so it is no wonder that the fundamental coincides with the left edge of the display. Notice that the Start frequency is 0 Hz and the Stop frequency is 3 GHz, twice the Center frequency. Using the number pad, enter in the Center frequency field 100 kHz, the frequency at the AFG output. This brings the fundamental to the center of the screen, so that it can be conveniently viewed.

But since it is a square wave, with near instantaneous rise and fall times, which are very high-frequency components, there should be lots of harmonics. Why are they not shown? The answer is in the Span. Since it is still 3 GHz with the interval between Center frequency and Stop being 1.50 GHz, the harmonics have diminished in amplitude to below the noise floor so that they are not visible. Using the number pad to set the Span to 10 m Hz, many harmonics are prominently displayed. The Center frequency, Start, and Stop are automatically adjusted to conform to the new Span.

If we increase Span, there will be an increased range of harmonics. Again, Start and Stop are automatically adjusted to conform to the new Span.

This is an introduction to operating a spectrum analyzer. In every new operation, it is necessary to go to the Freq/Span menu and configure the display to accommodate the input. There is much more to learn in order to become an expert spectrum analyzer user. The basics are in the User m anual, available free of charge (even if you have not yet purchased the instrument) from the manufacturer's website.

Bench-type spectrum analyzers provide a greater amount of spectral information than oscilloscopes, even in FFT mode. A drawback, however, is that they are substantially more costly. Also, there is the problem of displaying voltages that are referenced to but float above ground potential. Additionally, the high voltages encountered in three-phase, 480-volt equipment require special treatment from the standpoint of user and instrument.

As for voltage, inserting a step-down transformer or employing capacitive coupling has been suggested, but both of these solutions have the disadvantage that they will introduce additional, irrelevant harmonics. Practical solutions are to either use a 100:1 oscilloscope probe or to build a resistor divider with non-inductive resistors. Six of these high-watt components are needed for three-phase measurements. Both remedies are expensive if you don't already have the components, but tight tolerances are needed.

Turning to ripple on the DC bus, a hand-held, battery-powered oscilloscope set on AC coupling will make a valid measurement.

PC Based Spectrum Analyzer

For now, we'll go to the Tektronix RSA306B Real-Time Spectrum Analyzer. This instrument is a PC-based spectrum analyzer, which means that unlike the bench model, it connects to a user-supplied PC and takes advantage of its computing power and display capability, simplifying the instrument's hardware requirements. And yet together with your PC, the composite instrument has specifications and features comparable to far more expensive bench-top models. That is because nearly everyone owns or has access to a laptop or desktop PC.

The key ingredient in this mix is the software, which Tektronix furnishes in an included flash drive or as a download from the website. Once this software, SignalVu-PC, is installed in your computer, you can do a little simple cabling and then you're ready to access signals in the frequency domain and apply the instrument's powerful analytics to them.

The RSA306B comes with a Demo Three board, which provides an array of signals that demonstrate the capabilities of this PC-based spectrum analyzer. After the instrument's functions have been learned, the user is ready to disconnect the board and in its place hook up to some equipment to be investigated.

The board and module are powered by the PC. No separate AC connection is required. Included in the package is a specialized dual USB cable. It plugs into two USB slots in the PC. This is necessary due to the current requirements of the board. If you use a standard USB cable, you risk damaging the PC. The single end of this cable, with a Type B termination, plugs into the board.

Then, run the standard furnished USB cable from the module to the PC. Connect the BNC cable having the RF connector to the module and, for this demonstration, connect the other end of the BNC cable to the RF port on the board.

Now we're ready to use the PC-based spectrum analyzer to view basic spectrum measurements with markers. To do this, follow these steps:

1. Press the On button on the Demo Three board. The amber power light goes on, along with several green LEDs that indicate the status of the board.

2. On the board, set the reference to Internal.

3. Set the run mode to free run.

4. Click the Row and Column buttons to select CW as the signal to be generated.

On the RSA306B interface in the PC:

1. Double-click the SignalVu-PC icon on the PC desktop to start the application.

2. Click Live Link on the menu bar to view the drop-down menu. Click Search for Instrument. A notification will appear that the instrument has been found.

3. Click Connect to Instrument. Select RSA306 @ USB USB:00. First-time connection to the analyzer may take up to 10 seconds.

4. A Connect Status dialog box will appear to confirm that the instrument is connected. Click OK.

   You can verify connection status by looking at the connection indicator square on the menu bar. It is green when the instrument is connected, red when the instrument is not connected. You can also view the name of the instrument that is connected by hovering the mouse indicator over the green square.

5. Click the Preset button.

6. Click the Settings button to display the Settings control panel.

7. Click the m ax Span to set the span to the maximum value.

8. Right click the screen and select m arker to Peak. To right click on a tablet, touch the screen where you want to right click, hold until a complete square appears, then lift your finger. The shortcut menu appears.

9. Click to Center in the bottom m arker Settings area. This sets the center frequency of the analyzer to the center of the signal.

10. Set the Span to 40 m Hz.

11. Click Peak.

12. Click to Center again.

13. Right click the screen and select Add m arker.

14. Drag the new marker m 1 to the noise level.
    Notice that Readouts show:

   - The amplitude and frequency of the selected marker
   - The difference between the selected marker position and the position of the reference marker

15. Set the Resolution Bandwidth to 30 kHz. This RBW setting is located at the left side of the spectrum display. The RBW determines the Fast Fourier Transform bin size, or the smallest frequency that can be resolved. Smaller RBW improves the selectivity, but degrades the sweep speed and trace update rate.

Thermal Imager

An important diagnostic instrument, second only to the multimeter, is the thermal imager.

This hand-held, battery-powered instrument is aimed at the area of interest. In real time, the user sees in a remarkably clear display an image in which temperature gradations are represented in contrasting colors. In some models, areas of normal ambient temperature are shown in grayscale so that hot spots stand out.

First used by firefighters to find hidden flames, the thermal imager is ideal for checking motors, entrance panels, and all sorts of wiring and moving equipment for high-temperature areas, indicating potential failure. Poor electrical terminations are prominently displayed, so that they can be cleaned and retorqued, and hot bearings are indicated, so they can be lubricated or otherwise serviced, preventing an extended outage.

STUDY QUESTIONS

1. The simplest electrical test instrument is:
   A. a plug-in bulb socket
   B. a multimeter
   C. an oscilloscope
   D. a spectrum analyzer

2. The neon bulb tester can be used:
   A. on 120-volt circuits only
   B. on 120- and 240-volt circuits
   C. to measure voltage
   D. to measure resistance

3. A circuit analyzer:
   A. measures voltage in a circuit
   B. measures resistance in a circuit
   C. checks for common wiring faults
   D. costs more than a multimeter

4. The Fluke m odel 87V multimeter:
   A. cannot measure a VFD output
   B. can measure a VFD output
   C. has inputs protected to 10,000 volts
   D. will determine CAT ratings

5. The Fluke non-contact T6–600 electrical tester:
   A. reads volts but not amps
   B. reads amps but not volts
   C. requires probes to contact terminals
   D. can perform resistance, continuity, and diode tests

6. In diagnosing a three-phase motor, begin with the:
   A. spectrum analyzer
   B. oscilloscope
   C. phase rotation meter
   D. multimeter

7. The bench-type oscilloscope with standard probes is suitable for:
   A. measuring three-phase power quality
   B. measuring ripple on a VFD DC bus
   C. checking the input of an elevator's induction motor
   D. checking motion controller outputs

8. The Fluke 190-502 Scopem eter is suitable for:
   A. measuring three-phase power quality
   B. measuring ripple on a VFD DC bus
   C. checking the input at the induction motor
   D. all of the above

9. The Tektronix m DO3000 Series Oscilloscope in digital mode:
   A. displays logic high in blue and logic low in green
   B. displays logic high in green and logic low in blue
   C. displays a single transition during the time represented by one pixel column in red
   D. displays multiple transitions during the time represented by one pixel column in red

10. The Tektronix m DO3000 Series Oscilloscope can display up to _____ digital channels simultaneously.
   A. 10
   B. 12
   C. 14
   D. 16

For answers, go to Appendix A.

CHAPTER 7

MEASURING POWER QUALITY: HOW IT MAY BE IMPROVED

All electrical conductors undergo temperature rise when conveying current. If there is too much heat the conductor is damaged, so it is beneficial to limit current flow. Three-phase electrical distribution systems are characterized by reduced current flow given the amount of power delivered to the load. They do this by creating three separate phases with the load balanced among them. A circuit that is comprised of hot legs 120° out of phase with one another can deliver more power through smaller conductors. Galileo Ferraris, Mikhail Dolivo-Dobrovolsky, Jonas Wenström, and Nikola Tesla in the 1880s independently invented poly-phase systems. Tesla conceived and developed the three-phase distribution system and the three-phase induction motor. A three-phase waveform is shown in Figure 7-1.

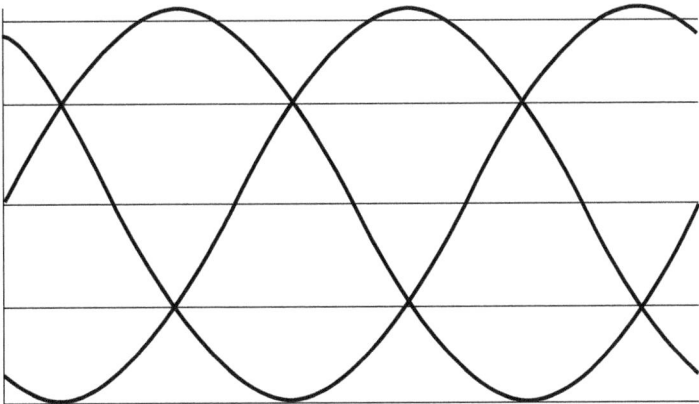

FIGURE 7-1 Ideal three-phase voltage waveforms—real ones would typically have superimposed noise.

Three-phase power is generated in either of two configurations, Y or delta. The generator has three windings, arranged symmetrically. The current in each winding is separated by the same phase angle, one-third cycle, from the other two. It can be expressed as 120° or $2\pi/3$ radians. Outside the generator, the current from each winding passes through one or more transformers, where the current and voltage, inversely related, are stepped up or stepped down without changing the phase spacing or frequency. At the output, a transformer converts the power to a specified level and supplies it through three wires to the point of connection.

Three-Phase Configurations

The more common Y-configuration connects one side of each winding to one of the three bus bars in the entrance panel and the other leg to a common, usually grounded, neutral bar. At the entrance panel, three-phase breakers connect to the three bus bars to supply three-phase loads, and single-pole breakers connect to one of the bus bars supplying single-phase loads. Thus, three-phase and single-phase power can be derived from a single entrance panel or load center without benefit of a transformer or phase converter, rotary, or electronic. Where phase-to-phase loads are to be supplied, double-pole breakers are used.

Three-phase configurations may be Y or delta, as shown in Figure 7-2. A delta- (named for the Greek letter delta, Δ) connected transformer winding is connected between two primary phases. In an open delta system, only two transformers are used, while in a closed delta system, there are three transformers, one for each phase. If one of the transformers fails or needs to be removed, the system will continue to function as an open delta system at 58% capacity.

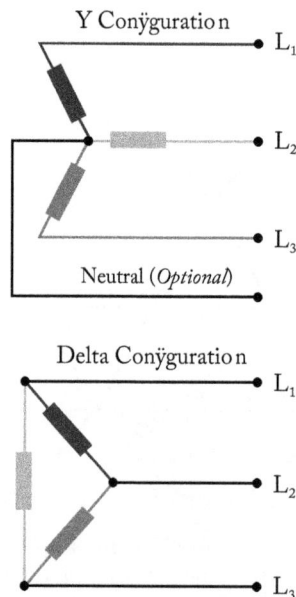

FIGURE 7-2 Y- and delta- configurations. (*Wikipedia*)

A double-pole breaker picks up the voltage between any two phases. A single-pole breaker picks up the voltage at one phase in conjunction with a neutral bar conductor. In either event, an equipment-grounding conductor should also be run to facilitate over-current operation.

In delta systems, ground connections made at the midpoints between two of the three phases are optional. These are known as "center grounded" three-phase delta systems. Because of the center tap, one of the three phases has a higher voltage with respect to ground than the other two. Care must be taken regarding this high leg. It is color-coded orange to distinguish it from the other two legs.

A three-phase motor is smaller, less expensive, and lasts longer than a single-phase motor of the same horsepower because it is vibration free and is required to dissipate less heat. For this reason, most induction motors over five horsepower are three-phase, although fractional horsepower three-phase motors are also available. Three-phase is easy to wire. Just run the three supply conductors, with over-current protection at the correct ampacity, to the motor and terminate them at the motor. Also run an equipment-grounding conductor and use a motor controller.

To reverse rotation, interchange two of the three lines. Some motor loads, such as fans or pumps, operate more efficiently running in one direction than the other. The blade or impeller shape is the reason. Correct rotation can be determined by trial and error, measuring the output. However, some pumps are instantly damaged by incorrect rotation.

Total Harmonic Distortion Measurements

Most electronics engineers and technicians have a good understanding of total harmonic distortion (THD). But there are a few elusive details that come into play during THD measurements. THD is the ratio of the sum of the powers of all harmonic components to the power of the fundamental frequency. Properly speaking, the fundamental frequency is the first harmonic, but THD discussions frequently don't acknowledge this fact. THD considers distortion contributed by second-order and higher harmonics but not by the random frequency, broad-spectrum distortion that is known as noise. THD plus noise is a separate though important metric.

The familiar sine wave, shown in Figure 7-3, is comprised of a single frequency, while non-sinusoidal waveforms are made up of two or more sine waves that can be added together on a point-by-point basis moving along the time-domain X-axis. Breaking down a complex non-sinusoidal waveform's sine wave components is a mathematically difficult process but became practical with the advent of the Fast Fourier Transform in the 1960s. Today, one simply imports the non-sinusoidal signal into a spectrum analyzer or, using Math Mode in an oscilloscope, presses FFT. Then, displayed on the screen in real time, is the signal at the channel input in the frequency domain, shown in Figure 7-4.

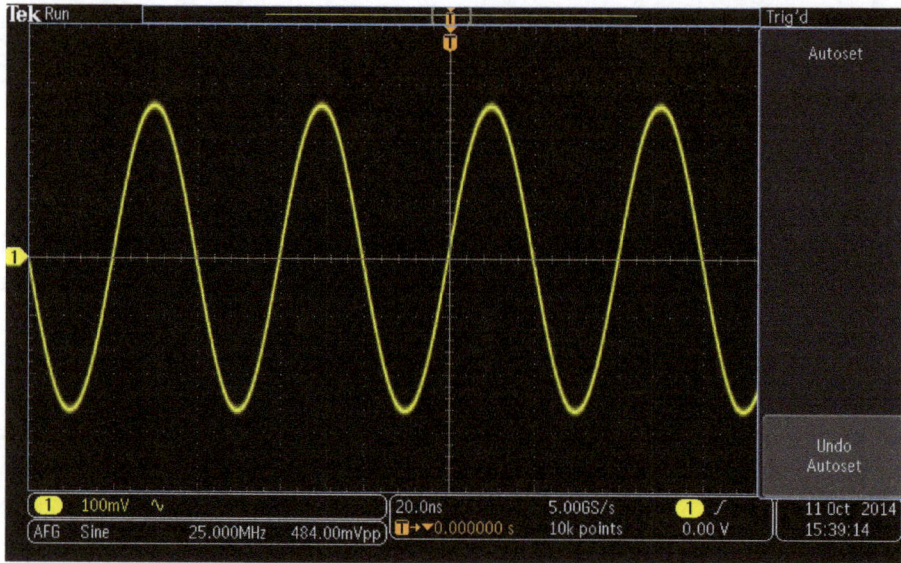

FIGURE 7-3   Sine wave in time domain.

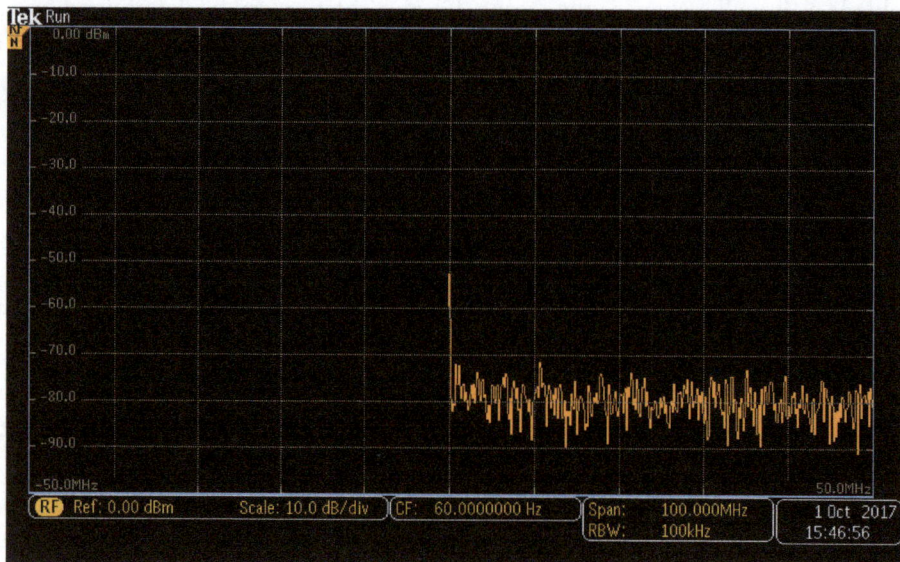

FIGURE 7-4   Sine wave in frequency domain.

Before scopes began to double as spectrum analyzers, the typical means of gauging THD was with a fundamental suppression THD analyzer. The instrument input is typically impedance-matched with the rejection circuit via an attenuator and an impedance matcher. This signal is then pre-amplified and sent to a Wien bridge notch filter tuned to reject the fundamental frequency and balanced for minimum output by adjusting the bridge controls. The output is the remaining signal after the fundamental has been suppressed. A feedback loop from the bridge amp output to the pre-amp input helps eliminate any remaining contribution from the fundamental frequency. The output from these blocks is measured, typically using an instrumentation amp driving an analog or digital meter. The voltage at the meter is caused by the amplitude, in units of power (dB) rather than volts. It displays on the Y-axis, and frequency, rather than time, displays along the Y-axis. These are the harmonics that, added together and divided by the fundamental, make up THD.

A high THD level in power systems is harmful for the system as well as for connected equipment. Lower THD equates to lower peak currents, higher efficiency, and higher power factor. Power factor is generally thought of as determined by the phase relationship between voltage and current, in accordance with: Power Factor $PF = \cos \theta_v - \cos \theta_i$, where θ_v is the phase angle of the voltage and θ_i is the phase angle of the current.

While this equation, known as the displacement factor, is valid when voltage and current are sinusoidal, it does not account for THD in non-sinusoidal circuits, which are prevalent today thanks to the rise of nonlinear loads with abundant harmonics.

Loads that include power conversion equipment — such as AC-DC, DC-AC and DC-DC, or nonlinear loads such as fluorescent ballasts — create a heavy nonlinear environment in which harmonics and THD abound. Switching power supplies, now common in office and home, contribute to this mix. This loading modifies the higher-quality sinusoidal power at the utility generator terminals. (Generators do contribute some fifth-order harmonics because of magnetic flux that takes place at the stator slots in addition to non-sinusoidal flux across the air gap.)

Scopes and THD meters aren't the only instruments capable of gauging harmonic content. Power analyzers, such as the PA3000 from Tektronix, shown in Figure 7-5, are optimized for characterizing power sources, including their harmonic content.

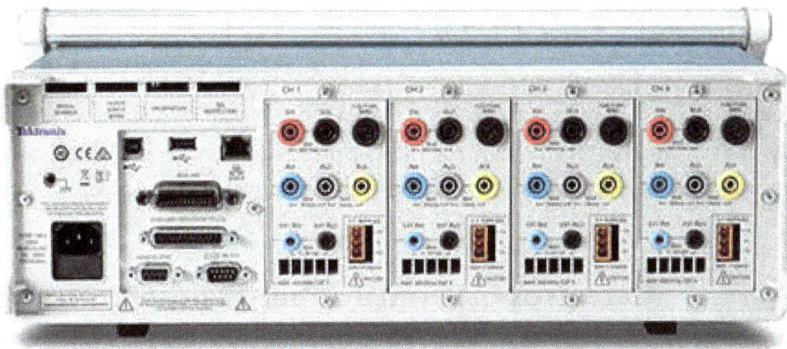

FIGURE 7-5 Tektronix PA3000 power analyzer. (*Tektronix*)

VFDs, welders, and arc furnaces also generate prodigious amounts of THD. Because harmonic currents are at higher frequencies than the power system fundamental, they see greater impedances. The cause of this strange phenomenon is that greater amounts of higher-frequency current flow near the surface of a conductor.

With less usable cross-sectional area, the effective resistance of the conductor rises, resulting in more heat. This is seen in three-phase neutral conductors and transformer windings.

When an AC motor is powered by a VFD, it gets a powerful direct dose of harmonics. This is a consequence of the high-speed switching in the VFD inverter section. Most of the ambient harmonics caused by other nonlinear loads in the same building or neighborhood are not much of a problem because they generally get suppressed when the power goes through the DC bus midway through the VFD. These outside harmonics do, however, assault the many autonomous motors that are found in the workplace.

For one thing, harmonics create flux distribution in motor air gaps, causing poor start-ups and abnormally high slip in induction motors. A serious problem in motors and generators is pulsating torque, causing losses and mechanical oscillations with harmful heat.

Here is the greatest problem in motors when there is high THD riding on the good power at the input: because of the alternating magnetic field, there is a normal temperature rise in the iron core due to eddy current and hysteresis loss. That is a given, and the iron core is by design sufficiently massive to dissipate this heat. But as it happens, the amount of eddy current loss varies with the square of the frequency. When high-frequency harmonics come along, the heat rises dramatically and, as it dissipates, a significant portion migrates into the windings, adding to the excess heat generated there by the unwanted harmonics and further stressing the winding insulation. Hysteresis varies directly with the frequency, not with its square, but still, it adds to the total.

Another factor, even more harmful, is a loss within the windings. This source of heat varies with the square of the current (I^2R), and the harmonics have a significant negative impact. Additionally, these high-frequency components exhibit harmful skin effect, reducing effective conductor size.

If a generator is to supply nonlinear loads, it should be derated because it has higher reactance and impedance than a similar size motor. Combined with high-frequency magnetic flux resulting from the presence of powerful harmonics, they boost the stator temperature. Rotor heating also results from these high-frequency currents.

Catastrophic Transformer Failure

Additionally, harmonics set the stage for often catastrophic transformer failure. Generally trouble-free, transformers may explode without warning as big nonlinear loads abruptly switch on. The problem is compounded in older transformers containing toxic PCB-laden cooling oil.

Copper and iron losses combine to create a hazardous situation. Eddy current rises when harmonics enter a transformer from line or load. Because eddy current is proportional to the square of the applied current and the square of its frequency, a transformer catastrophe can happen suddenly and without warning.

Harmonic current in transformers is a source of electromagnetic interference that can degrade nearby communication circuits. Shielding, increased spatial separation, and suppression of the harmonics are used to mitigate these effects.

To summarize, Fourier analysis (as opposed to Fourier synthesis) of a periodic signal reveals the harmonic frequencies that are components and integer multiples of the signal. This is where THD appears.

The reason that a voltage and its associated current are purely sinusoidal is that they consist of a single frequency. Multiple higher frequency components contribute to the observed THD. A square wave has a great amount of this distortion, while a sine wave that is in the real, non-ideal world has a small amount of it. In most cases, that component is not visible in the time domain, but it can usually be observed just above the noise floor in the frequency domain.

THD is a constant concern in power systems. Low power factor, higher peak currents, and low efficiency accompany high THD. In audio reproduction, a low THD equates to better fidelity. In communications systems, high THD means a potential for interference with nearby equipment and greater power consumption at the transmitter. Figure 7-6 shows a Keithley THD analyzer.

FIGURE 7-6   Keithley THD analyzer. (*Keithley*)

A THD analyzer can be used to measure the distortion of a waveform in comparison to a distortion-free sine wave. The instrument breaks the wave under investigation into its harmonics and compares each harmonic to the fundamental. An alternate procedure is to remove the fundamental by means of a notch filter, then to measure the remaining signal which will be the THD plus noise.

In audio equipment development, a low-distortion arbitrary function generator is used to insert an input into the unit being evaluated. Distortion at constituent frequencies is then measured for comparison of prototypes. In such procedures, crossover distortion for any given THD level is more audible and thus tends to outweigh clipping distortion, which produces higher-order harmonics. Generally, harmonics are beneficial only to the musician, who uses them in a flute or guitar to produce sounds that would otherwise be beyond the capability of the instrument.

The best way to mitigate harmonics is to suppress them at the source. An alternative is to create shielding or filters at the equipment that is affected by the harmonics. Then, measuring the amount of THD, the success of these measures can be evaluated.

FIGURE 7-7   Fluke Phase Rotation Indicator. (*Fluke*)

The Fluke phase rotation indicator, as shown in Figure 7-7, reveals wiring sequence for clockwise and counter-clockwise rotation. In a three-phase Y system or delta system without a grounded center tap in one of the windings, single-phase loads may be connected from a phase to neutral or across any two phases. This makes possible numerous single-phase voltages that can be used in various applications. If these loads are balanced, i.e., have equal impedance, transformers and conductors are used most economically.

In a balanced Y system, all three phase conductors have the same current and voltage with respect to the system neutral. With linear loads, the measured voltage between line conductors when the loads are equal is the square root of three times the phase-to-neutral voltage.

Costly Harmonics

A problem today is that an ever-increasing portion of connected loads are nonlinear. Ballasted fluorescent lighting, which is abundant in office settings, as well as switching power supplies and induction motors, are examples of non-linear loads. They produce costly third-order harmonics, which are in-phase in all three legs. As a result, they are additive in neutral conductors. This excess loading causes neutral heating in branch circuits and distribution lines all the way upstream including within utility generators.

Single-phase electronic loads generate harmonics at all multiples of the fundamental. The most harmful of these are triplen harmonics, because their amplitudes are highest. Higher-order harmonics diminish in amplitude as they become more remote from the fundamental as represented on the X-axis in an oscilloscope's frequency domain.

Three-phase loads do not generate triplen harmonics. Consequently, in industrial facilities with heavy three-phase loading, the greatest problem results from higher level odd harmonics—fifth, seventh, eleventh, and so on.

Active filters can mitigate harmonics, but they are complex and expensive to implement. They digitally synthesize reactive power to cancel the harmonics. A more economical solution is to use phase-shifting transformers to attenuate the harmonics. They work by combining harmonics from different sources that are phase-shifted with respect to one another so the harmonics then cancel out. Other harmonic mitigation techniques include the use of line reactors, harmonic traps, 12- and 18-pulse rectifiers, and low-pass filters.

Harmonics are also costly because they cause the apparent power in a system to exceed and stress the active and reactive components. Moreover, because they are higher frequency than the fundamental, harmonics reduce capacitive reactance, a parallel phenomenon, to a certain degree shunting out the intended load and heating up the supply wiring. In the presence of harmonics, capacitors see higher applied voltage, which can cause dielectric loss and actual damage. Three-phase induction motors also undergo loss and heating in their windings. Harmonics boost current and overheating in neutral conductors, which generally do not have over-current protection.

Power Factor

When large motors are not loaded to their full capacity, the cumulative effect within a facility is added to the presence of harmonics to reduce the power factor. Electric utilities frequently charge industrial customers a higher rate when power factor dips below 90%.

Power factor can be improved by adding power-factor-correction capacitors to the electrical system. A common implementation includes an automatic switch that places the capacitors on line only as needed.

Power factor correction capacitors require periodic inspection and maintenance. Thermal imaging is a good way to start. Workers should be aware that these devices are capable of storing lethal voltages long after they are powered down. Arc flash is a potential hazard as well. In that regard, anyone who works with test instruments in close proximity to three-phase power circuitry should wear personal protective equipment (PPE) as dictated by safety standards.

There are a couple of subtleties to be aware of when measuring three-phase electrical parameters. One concerns 480Y three-phase service. This configuration uses four wires, three hots, a neutral, and a ground wire. The voltage between any one leg and ground will be 277 V, and between any two hot wires you will measure 480 V. A transformer must be used to handle 120/208 single and three-phase loads. The transformer must have a 480Y primary and a 208Y secondary.

Three-phase machinery typically runs from delta voltage, a configuration using three hot wires and no neutral wire. If a 230-V machine is mistakenly connected to 480 V, its motor will most likely burn up. Voltage does not affect the RPM of a motor, but the frequency of the voltage does.

Finally, there are different ways of measuring three-phase power. Perhaps the simplest is to use a single power meter to measure the power in one phase at a time. The potential problem with this method is that it assumes the power in the unmeasured phases is the same as what's measured once the power meter is introduced in that phase.

The most straightforward method is to use a power meter in each phase leg simultaneously. Here, phase voltage for the power measurement is measured with respect to the neutral wire. Obviously, the total power is the sum of their readings. You can measure power accurately with only two power meters. One of the phases serves as the null reference and the power need only be measured for the remaining two phases. But there is a comparative calculation associated with this method that is employed as a check of its accuracy. It is easy to understand when the voltage source and load are in a Y configuration. Because the neutral isn't connected, the sum of the instantaneous currents in the three phases must be zero by Kirchhoff's current law: $I_1 + I_2 + I_3 = 0$.

It can then be demonstrated that the sum of the instantaneous powers of the three phases is equal to the instantaneous powers of two phases with the third phase (L2) as voltage reference:

$$V_1 \times I_1 + V_2 \times I_2 + V_3 \times I_3 = [(V_1 - V_2) \times I_1] + [(V_3 - V_2) \times I_3]$$

What's All This About Sine Waves?

Sine is one of the six trigonometric functions. (Where mathematicians speak of functions, electrical engineers say signals and scientists use the word phenomena. All mean the same thing.) Trigonometric functions are numbers assigned to various angles. These numbers equal the ratios of two sides of a right triangle, as shown in Figure 7-8.

The adjacent side is the line segment between angle theta and the right angle that defines that triangle. Opposite is the side opposite θ (Greek letter theta). Hypotenuse is the line segment between opposite and adjacent. Except in limiting cases, the hypotenuse is never longer than the opposite side or shorter than the adjacent side. Here are the ratios that equal trigonometric functions:

Sine = opposite/hypotenuse
Cosine = adjacent/hypotenuse
Tangent = opposite/adjacent
Cosecant = hypotenuse/opposite

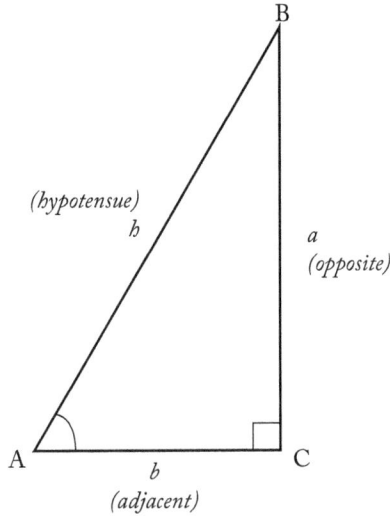

FIGURE 7-8 The three sides of a right triangle are hypotenuse, opposite, and adjacent.

Secant = hypotenuse/adjacent
Cotangent = adjacent/opposite

Since sine is the ratio of opposite over hypotenuse, for any given value of θ, it is always the same, regardless of the amplitude of the signal, which is represented by the length of the hypotenuse. The value of the trigonometric functions can be found in print or internet tables, or they can be obtained conveniently using a scientific calculator.

Students first learn to think of trigonometric functions in terms of triangles. But a better way is to look at them in the context of a circle. Many people correctly perceive that there is something basic or fundamental about the sine (and cosine) that is not applicable to other functions. Utility-supplied current world-wide, when it is not DC, is described by the sine wave, due to the rotary motion of a generator. If you look at the right triangle inscribed within a circle and visualize the way in which rotation affects the output of the generator, i.e., what happens as θ increases and the vector/hypotenuse traverses the complete circle, this equation will represent the state of affairs at any given instant:

$$V = A\sin 2\pi ft$$

where V is instantaneous voltage at the terminals
A is the amplitude
f is the frequency in hertz or cycles per second
t is time.

Why is it that the sine wave can be represented only by the sine function and the closely related cosine? Going back to the equation just cited, first let's clarify what 2 is doing in there. Sine is opposite over hypotenuse.

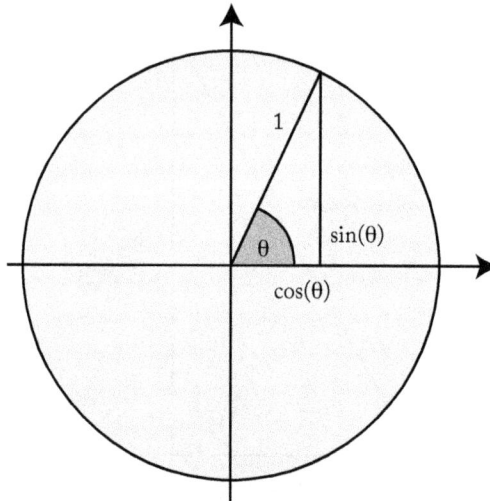

FIGURE 7-9   Right triangle inscribed in a circle. 1 is hypotenuse.

The hypotenuse, as shown in the inscribed triangle diagram, Figure 7-9, is the radius of that circle, whereas π refers to diameter, which is twice the radius. Hence 2 is a factor in the expression on the right side of the equal sign. *A*, amplitude, and *V*, voltage, are changing rapidly when a frequency in hertz of (let's say) 60 is factored in, never mind GHz, so we need to keep that in mind.

In your mind's eye, make the hypotenuse rotate (slowly, not 60 times per second!) and watch the ratio of opposite over hypotenuse change, amplitude and rate of change being inversely related. Ideally, this exercise would be done with the relevant trigonometric table at hand as well, but you get the idea.

The other very basic equation regarding the sine wave is:

$$V = A\sin\omega$$

where ω is angular frequency in radians. (A radian is a unit of angle, equal to an angle at the center of a circle whose arc is equal in length to the radius.) In the study of waveforms, it is beneficial to look at the sine and cosine functions in the context of a circle, and leave the right triangle to carpenters and surveyors.

If we contemplate the graph of a sine wave in the time domain, a couple of facts emerge. With amplitude in volts plotted on the Y-axis and time in seconds plotted on the X-axis, it is seen that the trace becomes vertical at 0 volts and it becomes horizontal at peak voltage, both of these for a vanishingly brief period to time. What this means is that the voltage is maximum when the rate of change is least and the voltage is minimum when the rate of change is greatest. These two relations bring to mind the behavior of capacitors, which block DC and pass high frequencies, and inductors, which block high frequencies and pass DC. All of this is true of DC except when it is being switched on or off, at which point it resembles a high-frequency AC due to the fast rise and fall times.

Another defining fact concerning the pure sine wave is that all the power is concentrated in the fundamental, i.e., all the power occurs at a single frequency. Another way of putting it is that if any power occurs outside the fundamental, the sine wave is not pure. Of course, a totally pure sine wave occurs only in an ideal world because any inductance in the load or generator would create a small harmonic content. Then too, there is the inevitable noise floor in the measuring instrument caused by thermal effects, not to mention mismatch in cabling characteristic impedance.

But these small harmonics can be disregarded in most applications. To evaluate power quality, we need to use an oscilloscope or specialized power-quality instrument. In the time domain, 60 Hz utility power will manifest in the oscilloscope display as the familiar sine wave. It is possible to get a rough idea of the power quality by looking at the trace. Large harmonics will show up as a slight wiggling of the curve. To really see what is going on, we have to perform a Fast Fourier Transform (FFT) analysis of the waveform.

To do this, in a Tektronix MDO3000 oscilloscope, with the sine wave displayed and adjusted to 60 Hz using Waveform Settings, press Math. Then press FFT. Press Menu Off once to get a clear view of the display. In a mixed-domain oscilloscope, the time domain and frequency domain can be plotted together against the same pair of axes. In the time domain, the Y-axis represents amplitude calibrated in volts. In the frequency domain, the Y-axis also represents amplitude, but it is calibrated logarithmically in decibels. In the time domain, the X-axis represents time, usually calibrated in seconds or milliseconds. In the frequency domain, the X-axis represents frequency. Because of the re-assignment of the X-axis, the frequency domain display has a totally different appearance in contrast to the time domain. It also serves different purposes, permitting the user to view the actual spectral distribution of the signal, rather than just gaining an impression of it as in the time domain.

To see the frequency domain display clearly, it is customary to alter the spectrum layout. To do this, in the frequency controls section on the instrument's front panel first press RF. Then press Freq/Span. Using multipurpose Knob A or, preferably, the number pad, change the center frequency to 60 Hz. This brings the fundamental to the center of the screen, making the display more readable. Additionally, you can enter different values for the span and watch the effect on the display. One hundred MHz is a good choice for viewing the frequency domain representation of a pure 60 Hz sine wave. Notice that as Span is changed, Start and Stop frequency adjust themselves accordingly.

Viewing a sine wave in the frequency domain, we see that all the power is concentrated in the fundamental, which is to say that there are no discernible harmonics. The irregular, low-amplitude trace outside the fundamental, across the entire spectrum, is the noise floor, present in all instruments due to thermal activity of the atoms in any device or conductor.

The time domain displays an accurate and highly intuitive graph of a signal's waveform shown as amplitude plotted against time. But to really see what is going on in terms of spectral analysis, it is the frequency domain that is more revealing.

Electrical equipment such as a variable frequency drive (which permits an operator on the factory floor or an elevator motion controller to regulate the speed and torque characteristics of an AC induction motor) requires a high-quality, usually three-phase, electrical supply. What is wanted is a pure sine wave, free of harmonics. The oscilloscope in FFT mode is ideal for this application. Since the voltage is high, it is essential to use the appropriate probes.

Other than an oscilloscope, a spectrum analyzer can be used to good effect to check power quality. This instrument is similar to the oscilloscope, with some important differences. The bench type spectrum analyzer is generally more expensive than an oscilloscope with similar specifications. It does not do a time-domain display, but it has more features in the frequency domain, so in this sense it is a more specialized instrument. To acquire a good display, center frequency and span are set as in the oscilloscope in FFT mode.

The oscilloscope and spectrum analyzer are both useful in seeing the true characteristics of a waveform, as opposed to merely measuring the voltage.

What you have to realize is that in a three-phase output from a generator, there are three separate windings equally spaced around the circle. They can be coming out of the inner spinning rotor or the outer stationary stator, it doesn't matter which. You have three pairs of wires, six in all. Then, inside the housing or outside, the six are connected in either a Y- or delta-configuration. Those are the three high-voltage lines that you see highest up on a power pole. They are stepped down in a three-phase transformer array to usable voltage, such as three-phase, Y-configuration, 480-volt.

What a lot of auto mechanics don't realize is that a regular automotive alternator generates three-phase power. It runs smoother with less vibration than would a single-phase unit. Inside or outside the housing, the six wires have in-line diodes, making DC, and these are paralleled up to make essentially a two-wire DC with a plus and a minus side.

There are many types of instruments that analyze power quality. Some are suitable for single phase only, while others are capable of three-phase measurement and display. Different voltage limits apply, and then there is the issue of fault current hazard in measuring voltages that are referenced to but float above the power supply ground plane.

The Fluke 434-II/435-II/437-II Three-Phase Energy and Power Quality Analyzer is a hand-held, battery-powered instrument with fully-insulated inputs that are rated 1000 volts RMS. It is powered by a rechargeable lithium-ion battery and an AC power adapter is included. The analyzer has a removable, upgradable 8 GH SD memory card, and comes with five test leads and alligator clips. It has a five-inch (diagonal) screen for displaying waveforms, readouts, and user interactions. A user manual can be downloaded free of charge from the manufacturer's website.

It's easy to get started with this user-friendly instrument, but to take full advantage of all measuring modes, the user manual is essential. Following six introductory chapters covering safety information, lithium-ion rechargeable battery use, summary of measuring modes, display information, and input connections, Chapters 7 through 22 contain explanations of measuring functions with tips and hints for getting the most out of the analyzer. The main topics are:

- Scope waveform and phasor
- Volts, amps, and hertz
- Dips and swells
- Harmonics
- Power and energy
- Energy loss calculator
- Power inverter efficiency
- Unbalance
- Inrush currents
- Power quality monitoring
- Flicker
- Transients
- Power wave
- Mains signaling
- Logger

Chapter 23 covers the use of cursors. Chapter 24, titled Setting Up the Analyzer, consists of an explanation of adjustments to customize measurements. Chapter 25 covers memory and the PC. It describes how to save, recall, and delete screenshots and data formats. Appendices contain information on principles of power measurement and energy loss calculation, installing USB drivers, and instrument security procedures.

Three-Phase Motors

Most motors rated over five horsepower, like the one shown in Figure 7-10, run on three-phase power, which has multiple advantages including higher efficiency and ease of wiring in contrast to single-phase. However, deficient power quality in a three-phase supply can set the stage for accelerated wear in an induction motor, if not immediate failure. Preventive maintenance and failure analysis invariably involve an examination of the three-phase power, measured at various points between the facility electrical service and the motor terminals.

The first instrument generally deployed is the familiar multimeter, which is suitable for measuring the all-important amplitude parameters. These metrics, while essential in determining whether the voltages in all three legs, with and without connected load, are substantially equal, do not constitute the complete power-quality picture. The oscilloscope, spectrum analyzer, energy logger, and power quality analyzer are all useful and have their place in shop and lab.

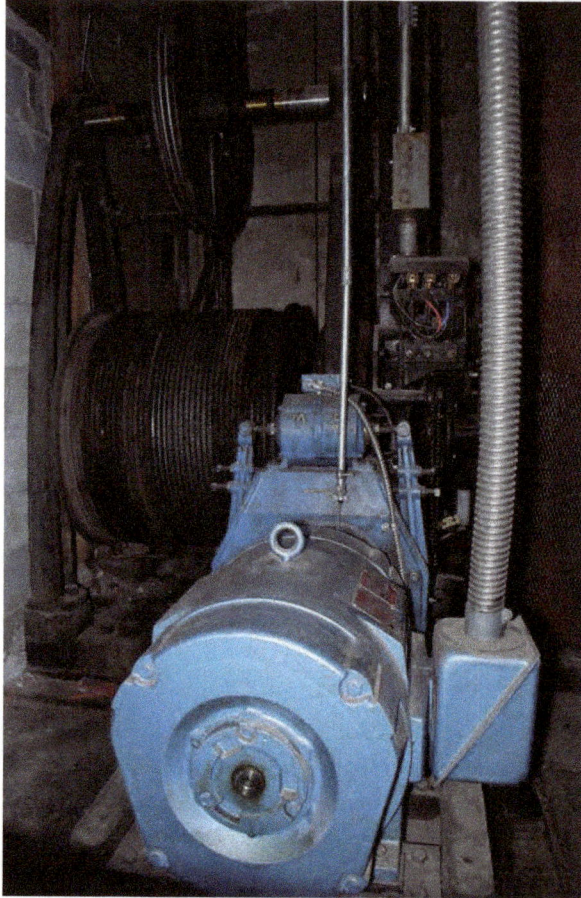

FIGURE 7-10   Three-phase motor powered by a variable-frequency drive.

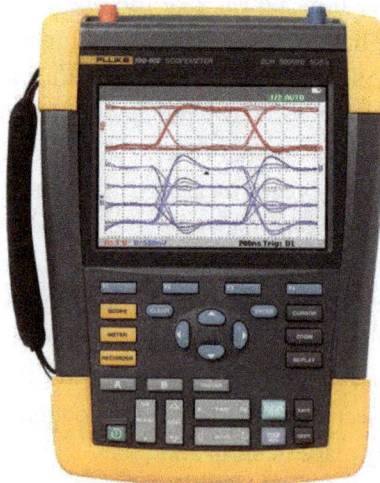

FIGURE 7-11   Four isolated inputs enable the Fluke Power Quality and Energy Analyzer
to safely display three-phase waveforms plus neutral simultaneously. (*Fluke*)

The Fluke 435 Series II Three-Phase Power Quality and Energy Analyzer, shown in Figure 7-11, sells on Amazon.com for just under $6,400. It measures and displays numerous details pertaining to the connected three-phase power. For example:

- Phase voltages are displayed and should be close to the utility nominal value and to the specified requirements for an installed VFD. The voltage waveforms must be a sine wave that is smooth and free of distortion. Scope Waveform displays the shapes of the three waveforms. Dips and Swells records sudden voltage changes. Transients Mode captures voltage anomalies.
- Phase currents: Volts/Amps/Hertz and Dips and Swells check current and voltage relations. Inrush current records sudden current inrushes such as motor inrush.
- Crest factor: If it is 1.8 or higher, there is waveform distortion.
- Harmonics mode checks for voltage and current harmonics and total harmonic distortion (THD) per phase. Trend records harmonics over time.
- Flicker is used to check long- and short-term voltage flicker and similar data in each phase. Trend records this data over time.
- Dips and Swells record sudden voltage changes down to one-half cycle.
- Frequency rarely differs from the utility nominal value. Volts/Amps/Hertz displays frequency. Use Trend Mode to search for frequency anomalies.
- Unbalance reveals phase voltage differences. They should be less than one percent of the average value of the three phases. Current difference in an individual phase must be less than ten percent.
- Energy Loss Calculator shows location and amount of energy losses.
- Power Inverter Efficiency displays the efficiency and energy output of connected inverters.
- Mains Signaling measures superimposed data present in the electrical supply.
- Power Wave is a high-resolution eight-channel waveform recorder.

Oscilloscope

Other than the multimeter, the oscilloscope is the most widely used diagnostic tool in electronics work, and this is particularly true in motor fault diagnosis. The instrument is suitable for checking three-phase power quality at the VFD input and ripple the DC bus. But you have to remember, as emphasized throughout the literature, that a conventional grounded bench-type oscilloscope, unless equipped with differential probes, cannot be used to display voltages where both sides of the circuit are referenced to but float above ground potential. In that hookup, the instant the probe ground reference lead contacts a floating voltage, hazardous current rushes through the circuit under investigation, ground reference lead, oscilloscope, and facility equipment grounding conductor, then back to the utility neutral at the entrance panel.

Power quality can be defined in a variety of ways, and sophisticated instrumentation is required to quantify it to an acceptable level. Most power quality issues reduce to voltage instability and waveform distortion. Since these properties vary over time as distributed loads go on- and offline, the instrumentation that monitors them must be capable of logging and storing power quality data over a period of time, often as long as a week, so that it can be accessed and analyzed as needed.

The ideal sine wave as provided by the utility may be distorted to varying degrees depending upon the connected loads. Utilities are generally OK with straight resistive loads, unless they are of so little impedance that they overload the utility/distribution system ability to power the load. With today's huge grids and robust distribution systems, that is not likely to happen except in the event of an electrical fault, whereupon redundant over-current devices quickly isolate the impacted line. Sensitive equipment, such as a motor already loaded to capacity, may respond to waveform clipping or flattening at peak amplitude by exhibiting a small temperature rise or outright overheating. After an extended period of time, winding insulation may deteriorate and at the very least there will be reduced motor life.

When the load is non-linear, temperature rise can be rapid and unrestrained. Assuming over-current protection is in place, the motor will trip out, causing an inconvenient outage.

Non-linear loads cause overheating as a consequence of ohm's law. Since $I = V/R$, where I is current in amps, V is voltage in volts and R is resistance in ohms, when resistance is constant, I and V vary in a linear fashion. The graph is a straight line. When the voltage is a sine wave, the current is also a sine wave.

Incandescent lamps and resistive heaters are examples of linear loads. Today, non-linear loads have become commonplace. They include computers, VFDs, fluorescent ballasts, switching power supplies, and uninterruptable power supplies. What is different about them is that the impedance varies periodically with the applied voltage. Using a current probe, you can examine the current in an oscilloscope display, and appropriately scaled, the waveform, while it may follow the overall form of a sine wave, will have irregularities that profoundly affect its appearance. In a spectrum analyzer (or oscilloscope in the FFT or RF mode), in addition to the single spike at the fundamental frequency, numerous harmonics will appear at regular intervals, diminishing in amplitude as the frequency difference increases.

Non-linear loads modify the applied voltage in a variety of ways. A solid-state power supply consists of semiconductors that abruptly begin to conduct only when the applied voltage rises to a specific level, which we may think of as the "firing voltage." It is at this point that charge carriers, in an NPN or PNP transistor, accumulate at the junction of the N and P layers so that the device more or less abruptly begins to conduct current. Insulated-gate bipolar transistors function according to different principles, but in terms of waveform distortion the end result is the same. The impedance of the device changes in the course of each applied sine wave. The results can be seen in an oscilloscope.

If the phases of the three sine waves in the applied voltages at a three-phase power supply differ in amplitude, the timing will be disrupted because firing voltages are not simultaneous. The DC will fluctuate beyond the ability of the filter components to control it. In a VFD system, this will ultimately translate to temperature rise in the motor. That is why it is important to examine the DC bus, using a multimeter or hand-held, battery-powered oscilloscope switched to AC coupling. (Observe voltage limits in all instruments, especially on the DC bus where voltage exceeds AC system voltage by a factor of 1.414. That is because in a full-wave rectifier, it is the peak-to-peak voltage of the AC input that is relevant.) In a three-phase Y-configured system, since the phases are 120 degrees apart, the currents in the neutral cancel out so that the neutral conductor actually carries no current. However, when connected to a non-linear load, this cancelling out does not take place uniformly throughout the waveform cycle. When harmonics combine on the neutral conductor, rather than cancel, they add, so that there can be more current on the neutral than on the phase conductors, and the over-current protection is not effective.

Heating of the neutral can take place to a greater or lesser extent depending upon the nature of the non-linear load. The greatest amount of heating takes place when there are strong triplen (third) harmonics because they are odd, which do not cancel, and also because they are higher amplitude than other odd harmonics, which rapidly diminish in amplitude as they are farther separated in frequency from the fundamental. (By convention, the fundamental is known as the first harmonic.) One remedy for this heating problem is to oversize the neutral.

Other ways to deal with hot neutrals are to use separate neutrals for each phase and to reduce (where possible) use of non-linear loads, for example by turning off some fluorescent lighting where not needed.

Fourier Transform

As far as we know, everything that exists, including the entire universe, can be described as a waveform, because everything is a function of a unique variable. Regardless of complexity, moreover, all waveforms are the sum of sine waves of different frequencies.

The Fourier Transform is a mathematical protocol that decomposes any waveform into its constituent sine waves. The Fourier Transform is universally applicable for all functions. (Remember that a function is the same as a waveform, just seen from a different perspective.) The Fourier Series is applicable to those functions that are periodic. To begin, we need to understand the meaning of periodic. We all know that something is periodic if it occurs over and over again without changing. More rigorously, a function is periodic, with period T, if:

$$F(t) = f(T + t)$$

is true for all t.

The mathematics required to perform the Fourier Transform on a periodic function is very difficult and time consuming, but the process was made far simpler by the rediscovery and refinement of the Fast Fourier Transform (FFT) in the 1960s. FFT, as its name implies, facilitates the display of a waveform in the frequency domain, in which the sinusoidal components of a complex non-sinusoidal waveform can be viewed.

Why be bothered? In the frequency domain, the spectral components of a waveform can be seen, whereas in the time domain it is the end result that is seen. Accordingly, particularly in the evaluation of power quality where we are looking for a pure waveform, the frequency domain shows the problem(s) and from there solutions may be found.

Modern digital storage oscilloscopes have this functionality. To see it in action, connect 120-volt utility power to an oscilloscope input, taking care to observe voltage limits and floating voltages. (A residential low-voltage thermostat transformer is helpful.) You should see a well-formed sine wave with no irregularities in the trace. Now press Math > FFT. The same signal is shown in the frequency domain. A mixed domain oscilloscope (MDO) displays the signals in both domains simultaneously, which is all the better. You may have to adjust Frequency/Span to display harmonics, if any. Turn off any computers or fluorescent lighting. You should see only the fundamental at 60 Hz and the omnipresent noise floor, with no strong harmonics.

To introduce harmonics, plug a hand-held electric drill with a universal motor into the other side of the duplex receptacle. When you press the trigger, you should see waveform distortion. If it looks like regularly-spaced small sinusoidal excursions of the utility waveform, you are seeing harmonics.

Examine the signal in the frequency domain. Harmonics appear as regularly-spaced spikes diminishing in amplitude as the spectral distance from the fundamental increases.

Noise is a broad-band phenomenon that does not diminish until you approach the bandwidth limitation of the instrument.

Try the same experiment with the drill plugged into a different branch circuit, and then a branch circuit on the opposite leg in the entrance panel.

STUDY QUESTIONS

1. Three-phase AC systems have three legs that are _____ degrees out of phase.
   A. 60
   B. 120
   C. 180
   D. 270

2. Three-phase systems are _____ configured.
   A. Y
   B. delta
   C. Y or delta
   D. AC or DC

3. 120 degrees is the same as:
   A. 2 π/3 radians
   B. 3 π/2 radians
   C. ¼ cycle
   D. ¾ cycle

4. To derive single-phase from three-phase:
   A. a phase converter is required
   B. a transformer is required
   C. a one- or two-pole circuit breaker is required
   D. a resister network is used

5. In a delta-configured three-phase system:
   A. a ground connection may be made at the center point between two phases
   B. the midpoint of one phase may be grounded
   C. one leg may be at a higher voltage to ground than the other two legs
   D. all of the above are true

6. Given the same horsepower, a three-phase motor is _____ compared to a single-phase motor.
   A. less expensive
   B. longer lasting
   C. more vibration free
   D. all of the above

7. To reverse rotation in a three-phase motor:
   A. interchange any two leads
   B. roll over all three connections
   C. rewire it to single-phase
   D. a VFD is needed

8. Total harmonic distortion:
   A. is caused by motor vibration
   B. is the sum of all harmonic components
   C. can be harnessed to get more power out of a motor
   D. is generally harmless

9. Total harmonic distortion can be displayed:
   A. in a multimeter in frequency mode
   B. in an oscilloscope in time domain
   C. in an oscilloscope in frequency domain
   D. using a circuit analyzer

10. Total harmonic distortion is generated by:
   A. VFDs
   B. welders
   C. arc furnaces
   D. any of the above

For answers, go to Appendix A.

CHAPTER 8

ELEVATOR SYSTEM OVERVIEW

An elevator installation, whether it consists of a single older car and hoistway in a four-story building or a group installation with computers governing multiple motion controllers in a large high-rise, is best seen as a complete system. Mechanical, structural, and electrical parts and above all, the motion controller, as shown in Figure 8-1, must function in concert if the complete system is to work safely and efficiently.

Technicians and maintenance workers have to go beyond responding to the occasional outage, which can sometimes be fixed in a few minutes while other times a joint effort measured in hours is required.

After the elevator system is back in service, the technicians should analyze the event and decide if further action is needed to ensure that the malfunction does not recur tomorrow or next year.

An important element, sometimes neglected, is scheduled preventive maintenance. While doing their rounds, maintenance workers should continually scrutinize and evaluate the entire installation with a view to detecting any flawed or missing elements that could give rise to hazards down the road. Technicians and support workers in the elevator trade should be familiar with applicable codes and also exercise abundant common sense. While most installations incorporate redundant safeguards, sometimes hidden design flaws conspire to create unexpected hazards so that workers and members of the public are at risk.

Realistically, we have to contemplate some rare but awful scenarios. For example, an unsupervised, exuberant youngster jumps in an elevator car and, seeing that it resembles a trampoline, jumps up and down in a regular rhythm, causing the steel rope, as shown in Figure 8-2, to jump off a non-compliant top sheave.

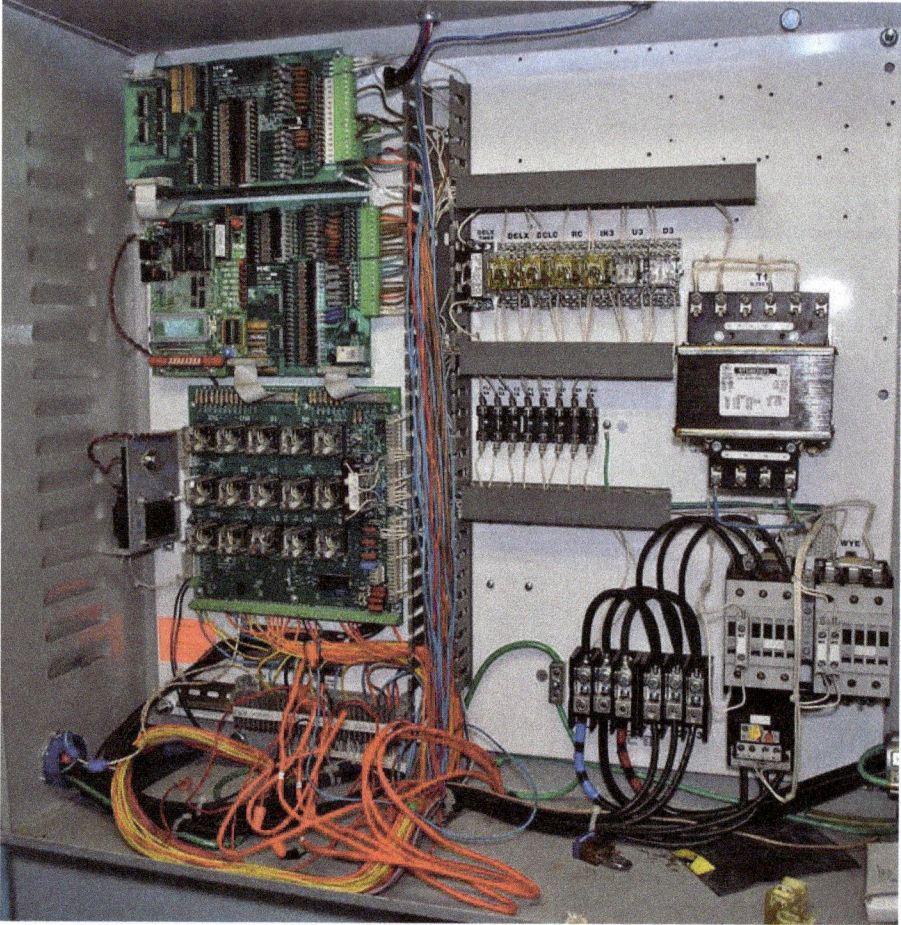

FIGURE 8-1 Elevator motion controller. (*Judith Howcroft*)

FIGURE 8-2 Elevator steel rope. (*Renown Electric*)

The car drops, smashing to pieces in the pit below the hoistway, killing the boy, only because maintenance workers lubricated the rails with a grease not compatible with the safeties despite warnings in the service manual. Or: Rebuilding a printed circuit board in a motion controller (this really happened), two wires were reversed and connected to the wrong lugs. This disabled a car door interlock, decapitating a young passenger.

Elevator fatalities are very rare, and most involve maintenance workers who step through an open door into a darkened hoistway. These events, despite their rarity, are very tragic for the people involved and for their families. Beware of an unlikely combination of circumstances that gets around our redundant safeguards.

Preventive maintenance is usually thought of as lubricating moving parts according to a time schedule, either calendar-based or referenced to operating hours. This is specified in the user manual, along with other types of preventive maintenance and inspections that should be performed. But conscientious workers, when making their rounds, scrutinize the entire installation, looking for any non-compliant or damaged components that require upgrading. To succeed in this, plain common sense is needed in addition to a thorough familiarity with applicable codes, particularly ASME A17.3, Safety Code for Existing Elevators and Escalators and the National Electrical Code.

National Electrical Code

NEC requirements relate specifically to electrical safety issues. These are the twin demons of electrical shock and electrical fire, although there are some additional concerns. For example, an inadequately secured and supported length of metal conduit could fall, striking a passerby. But primarily it is shock and fire that are the safety issues.

In addition to the entire NEC, Article 620 addresses elevators and related equipment—dumbwaiters, escalators, moving walks, platform lifts, and stairway chairlifts. NEC does not cover internal components within factory-made equipment, typically with housings or enclosures, such as electric motors, light fixtures, and motion controllers. Items of this sort are listed and labeled by third-party testing organizations, notably Underwriters Laboratories (UL).

NEC, first promulgated in the 1800s, is currently revised and amended in three-year cycles. Public input is solicited, proposed revisions are acted upon by committees of experts in the individual fields, and the entire text is voted upon at the NFPA convention.

Revisions are quite extensive. Many states and jurisdictions do not adopt the current edition until long after NFPA releases it. In some areas, 2008, 2011, and 2014 NEC are still enforced as of 2020, so to avoid installation conflicts and before taking a licensing exam it is essential to check with the local electricians' licensing board.

Licensing is not covered in the NEC. It is administered by the local jurisdiction. In working on electrical systems, it is important to comply with statutory electrical requirements, and in the United States NEC is the applicable reference. Electricians are very familiar with this hefty volume, and when working on elevator systems, they invariably refer to the entire work as well as Article 620. In this chapter, rather than discussing every detail, we will review the fundamental concepts.

Section 620.2, Definitions, states that the motor controller, motion controller, and operation controller are located in a single enclosure or a combination of enclosures. These are some of the terms that are defined:

- **Control Room.** An enclosed control space outside the hoistway, intended for full bodily entry, that contains the elevator motor controller. The room could also contain electrical and/or mechanical equipment used directly in connection with the elevator but not the electric driving machine or the hydraulic machine.
- **Control Space** (for elevator). A space inside or outside the hoistway, intended to be accessed with or without full bodily entry, that contains the elevator motor controller. This space could also contain electrical and/or mechanical equipment used directly in connection with the elevator, but not the electric driving machine or the hydraulic machine.
- **Control System.** The overall system governing the starting, stopping, direction of motion, acceleration, speed, and retardation of the moving member.
- **Controller, Motion.** The electrical device(s) for that part of the control system that governs the acceleration, speed, retardation, and stopping of the moving member.
- **Controller, Motor.** The operative units of the control system comprised of the starter device(s) and power conversion equipment used to drive an electric motor, or the pumping unit used to power hydraulic control equipment.
- **Controller, Operation.** The electrical device(s) for that part of the control system that initiates the starting, stopping, and direction of motion in response to a signal from an operating device.
- **Machine Room** (for elevator). An enclosed machinery space outside the hoistway, intended for full bodily entry, that contains the electrical driving machine or the hydraulic machine. The room could also contain electrical and/or mechanical equipment used directly in connection with the elevator.
- **Machinery Space** (for elevator). A space inside or outside the hoistway, intended to be accessed with or without full bodily entry, that contains elevator mechanical equipment, and could also contain electrical equipment used directly in connection with the elevator. This space could also contain the electrical driving machine or the hydraulic machine.
- **Operating Device.** The car switch, pushbuttons, key or toggle switch(s), or other devices used to activate the operation controller.
- **Remote Machine Room and Control Room.** A machine room or control room that is not attached to the outside perimeter or surface of the walls, ceiling, or floor of the hoistway.

- **Remote Machinery Space and Control Space.** A machinery space or control space that is not within the hoistway, machine room, or control room and that is not attached to the outside perimeter or surface of the walls, ceiling, or floor of the hoistway.
- **Signal Equipment.** Includes audible and visual equipment such as chimes, gongs, lights, and displays that convey information to the user.

Section 620.3, Voltage Limitations, provides that the supply voltage is not to exceed 300 volts between conductors unless otherwise permitted in Section 620.3(A) through (C).

A. Power Circuits provides that branch circuits to door operator controllers, door motors, and branch circuits, and feeders to motor controllers, driving machine motors, machine brakes, and motor-generator sets are not to have a circuit voltage in excess of 1000 volts. Internal voltages of power conversion equipment and functionally associated equipment, and the operating voltages of wiring interconnecting the equipment are permitted to be higher, provided that all such equipment and wiring is listed for the higher voltages. Where the voltage exceeds 600 volts, warning labels or signs that read "DANGER – HIGH VOLTAGE" are to be attached to the equipment and be plainly visible.

B. Lighting circuits are to comply with the requirements of Article 410.

C. Branch circuits for heating and air-conditioning equipment located on the elevator car are not to have a circuit voltage in excess of 1000 volts.

620.4, Live Parts Enclosed, provides that all live parts of electrical apparatus in the hoistways, at the landings, and in or on the cars of elevators are to be enclosed to protect against accidental contact.

620.5, Working Clearances, states that working clearances are to be provided about controllers, disconnecting means, and other electrical equipment in accordance with 110.26(A).

Part II. Conductors

Section 620.11 provides that the insulation of conductors is to comply with (A) through (D).

A. The conductors to the hoistway door interlocks from the hoistway riser are to be one of the following:

1. Flame retardant and suitable for a temperature of not less than 392°F. Conductors are to be Type SF or equivalent.

2. Physically protected using an approved method such that the conductor assembly is flame retardant and suitable for a temperature of not less than 392°F.

B. Traveling cables used as flexible connections between the elevator car or counterweight and the raceway are to be of the types of elevator cables listed in Table 400.4 or other approved types.

C. Other wiring: All conductors in raceways are to have flame-retardant insulation.

D. Insulation: All conductors are to have an insulation voltage rating equal to at least the maximum nominal circuit voltage applied to any conductor within the enclosure, cable, or raceway. Insulations and outer coverings that are marked for limited smoke and are so listed are permitted.

620.12 provides that the minimum size of conductors, other than conductors that form an integral part of control equipment are to be in accordance with (A) and (B).

A. Traveling cables:

1. In lighting circuits, 14 AWG copper. 20 AWG copper or larger conductors are permitted in parallel, provided the ampacity is equivalent to at least that of 14 AWG copper.

2. In other circuits, 20 AWG copper.

B. Other wiring—24 AWG copper. Smaller size listed conductors are permitted.

620.13, Feeder and Branch-Circuit Conductors. Conductors are to have an ampacity in accordance with 620.13 (A) through (D). With generator field control the conductor ampacity is to be based on the nameplate current rating of the driving motor or the motor-generator set that supplies power to the elevator motor.

A. Conductors supplying a single motor are to have an ampacity not less than the percentage of motor nameplate current determined from 430.22 (A) and (E).

NEC Article 620 contains this Informational Note, which clarifies the ampacity requirement based on an elevator's inherent duty cycle, because it frequently rests when stopped at landings.

Informational Note: Some elevator motor currents, or those motor currents of similar function, exceed motor nameplate value. Heating of the motor and conductors is dependent on the root-mean square (RMS) current value and the length of operation time. Because this motor application is inherently intermittent duty, conductors are sized for duty cycle service as shown in Table 430.22(E).

B. Conductors supplying a single motor controller are to have an ampacity not less than the motor controller nameplate current rating, plus all other connected loads. Motor controller nameplate current ratings are permitted to be derived based on the RMS value of the motor current using an intermittent duty cycle and other control system loads, if present.

C. Conductors supplying a single power transformer are to have an ampacity not less than the nameplate current rating of the power transformer plus all other connected loads.

An Informational Note states that the nameplate current rating of a power transformer supplying a motor controller reflects the nameplate current rating of the motor controller at line voltage (transformer primary).

D. Conductors supplying more than one motor, motor controller, or power transformer are to have an ampacity not less than the sum of the nameplate current ratings of the equipment plus all other connected loads. The ampere ratings of motors to be used in the summation are to be determined from Tables 430.22(E), 430.24, and 430.24 Exception No. 1.

620.14, Feeder Demand Factor, provides that feeder conductors of less ampacity than required by 620.13 are permitted, subject to the requirements of Table 620.14.

620.15, states that the motor controller rating is to comply with 430.83. The rating is permitted to be less than the nominal rating of the elevator motor when the controller inherently limits the available power to the motor and is marked as power limited.

620.16, Short-Circuit Current Rating, provides:

A. Where an elevator control panel is installed, it is to be marked with its short-circuit current rating, based on one of the following:

1. Short-circuit current rating of a listed assembly.

2. Short-circuit current rating established using an approved method.

B. Installation. The elevator control panel is not to be installed where the available short-circuit current exceeds its short-circuit current rating, as marked in accordance with 620.16(A).

Part III. Wiring

620.21, Wiring Methods, states that conductors and optical fibers located in hoistways, machinery spaces, control spaces, in or on cars, in machine rooms and control rooms, not including the traveling cables connecting the car or counterweight and hoistway wiring, are to be installed in rigid metal conduit, intermediate metal conduit, electrical metallic tubing (as shown in Figure 8-3), rigid nonmetallic conduit or wireways, or is to be Type MC, MI or AC cable unless otherwise permitted in 620.21(A) through (C).

FIGURE 8-3 Electrical metallic tubing (EMT). (*Judith Howcroft*)

Exception: Cords and cables of listed cord- and plug-connected equipment is not to be installed in a raceway.

Elevators

1. Hoistways and Pits

   a. Cables used in Class 2 power-limited circuits are permitted, provided the cables are supported and protected from physical damage and are of a jacketed and flame-retardant type.

   b. Flexible cords and cables that are components of listed equipment and used in circuits operating at 30 volts RMS or less or 42 volts DC or less are permitted provided the cords and cables are supported and protected from physical damage and are of a jacketed and flame-retardant type.

   c. The following wiring methods are permitted in the hoistways in lengths not to exceed six feet:

      1. Flexible metal conduit

      2. Liquidtight metal conduit

      3. Liquidtight flexible metallic conduit

      4. Flexible cords and cables, or conductors grouped together and taped are permitted to be installed without a raceway. They are to be located so as to be protected from physical damage and are to be of a flame-retardant type and are to be part of the following:

         a. Listed equipment

b. A driving machine

c. A driving machine brake

Exception: The conduit length is not required to be limited between risers and limit switches, interlocks, operating buttons, and similar devices.

d. A sump pump or oil recovery pump in the pit is permitted to be cord connected. The cord is to be hard usage oil-resistant type, of a length not to exceed six feet, and is to be located so as to be protected from physical damage.

5. Cars

a. Flexible metal conduit, liquidtight flexible metal conduit or liquidtight flexible nonmetallic conduit of metric designator 12 (trade size 3/8) or larger, not exceeding six feet in length is permitted on cars so located as to be free from oil and securely fastened in place.

Exception: The conduit length is not required to be limited between risers and limit switches, interlocks, operating buttons, and similar devices.

b. Hard-service cords and junior hard-service cords that conform to the requirement of Article 400 Table 400.4 are permitted as flexible connections between the fixed wiring on the car and devices on the car doors or gates. Hard-service cords only are permitted as flexible connections for the top-of-car operating device or the car-top work light. Devices or luminaires are to be grounded by means of an equipment grounding conductor run with the circuit conductors. Cables with smaller conductors and other types and thicknesses of insulation and jackets are permitted as flexible connections between the fixed wiring on the car and devices on the car doors or gates, if listed for this use.

c. Flexible cords and cables that are components of listed equipment and used in circuits operating at 30 volts RMS or less or 42 volts DC or less are permitted, provided the cords and cables are supported and protected from physical damage and are of a jacketed and flame-retardant type.

d. The following wiring methods are permitted on the car assembly in lengths not to exceed six feet:

1. Flexible metal conduit

2. Liquidtight flexible metal conduit

3. Liquidtight flexible nonmetallic conduit

4. Flexible cords and cables, or conductors grouped together or corded are permitted to be installed without a raceway. They are to be located so as to be protected from physical damage and are to be of a flame-retardant type and are to be part of the following:

a. Listed equipment

b. A driving machine; or

c. A driving machine brake

5. Within machine rooms, control rooms, and machinery spaces and control spaces:

a. Flexible metal conduit, liquidtight flexible metal conduit or liquidtight flexible nonmetallic conduit of metric designator 12 (trade size 3/8) or larger, not exceeding six feet in length, is permitted between control panels and machine motors, machine brakes, motor-generator sets, disconnecting means, and pumping unit motors and valves.

   An exception is that liquidtight flexible nonmetallic conduit, trade size ⅜ or larger, is permitted to be installed in lengths in excess of six feet.

b. Where motor-generators, machine motors or pumping unit motors and valves are located adjacent to or underneath control equipment and are provided with extra-length terminal leads not exceeding six feet in length, such leads are permitted to be extended to connect directly to controller terminal studs without regard to the carrying-capacity requirements of Articles 430 and 445. Auxiliary gutters are permitted in machine and control rooms between controllers, starters, and similar apparatus.

c. Flexible cords and cables that are components of listed equipment used in circuits operating at 30 volts RMS or 42 volts DC are permitted, provided the cords and cables are supported and protected from physical damage and are of a jacketed and flame-retardant type.

d. On existing or listed equipment, conductors are permitted to be grouped together and taped or corded without being installed in a raceway. Such cable groups are to be supported at intervals not over three feet and located so as to be protected from physical damage.

e. Flexible cords and cables in lengths not to exceed six feet that are of a flame-retardant type and located so as to be protected from physical damage are permitted in these rooms and spaces without being installed in a raceway. They are to be part of the following:

   1. Listed equipment

   2. A driving machine; or

   3. A driving machine brake

4. Counterweight. The following wiring methods are permitted on the counterweight assembly in lengths not to exceed six feet:

a. Flexible metal conduit

b. Liquidtight flexible metal conduit

c. Liquidtight flexible nonmetallic conduit

d. Flexible cords and cables or conductors grouped together and taped or corded are permitted to be installed without a raceway. They are to be protected from physical damage, are to be of a flame-retardant type, and are to be part of the following:

1. Listed equipment

2. A driving machine; or

3. A driving machine brake

Part IV. Installation of Conductors

620.32, Metal Wireways and Nonmetallic Wireways, states that the sum of the cross-sectional areas of the individual conductors in a wireway is not to be more than 50 percent of the interior cross-sectional area of the wireway.

Vertical runs of wireways are to be securely supported at intervals not exceeding 15 feet and are to have not more than one joint between supports. Adjoining wireway sections are to be securely fastened together to provide a rigid joint.

620.33, Number of Conductors in Raceways. The sum of the cross-sectional areas of the individual conductors in raceways is not to exceed 40 percent of the interior cross-sectional area of the raceway, except as permitted in 620.32 for wireways.

620.34, Supports for Cables or Raceways in a Hoistway, states that they are to be securely fastened to a guide rail.

620.35, Auxiliary Gutters, states that they are not subject to the restrictions of 366.12(2) covering length or of 366.22 covering number of conductors.

620.36, Different Systems in One Raceway or Traveling Cable. Optical fiber cablers and conductors for operating devices, operation and motion control, signaling, fire alarm, lighting, heating, and air-conditioning circuits of 1000 volts are less are permitted to be run in the same traveling cable or raceway system if all conductors are insulated for the maximum voltage applied to any conductor within the cables or raceway system and if all live parts of the equipment are insulated from ground for this maximum voltage. Such a traveling cable or raceway is also permitted to include shielded conductors and/or one or more coaxial cables if such conductors are insulated for the maximum voltage applied to any conductor within the cable or raceway system. Conductors are permitted to be covered with suitable shielding for telephone, audio, video, or higher frequency communication circuits.

620.37, Wiring in Hoistways, Machine Rooms, Machinery Spaces and Control Spaces.

A. **Uses Permitted.** Only such electrical wiring, raceways, and cables used directly in connection with the elevator, including wiring for signals, for communication with the car, for lighting, heating, air-conditioning and ventilating the elevator car, for fire detection systems, for pit sump pumps, and for heating, lighting, and ventilating the hoistway, machine rooms, control rooms, machinery spaces and control spaces.

B. **Lightning Protection.** Bonding of elevator rails (car and/or counterweight) to a lightning protection system down conductor is permitted. The lightning protection system down conductor(s) is not to be located within the hoistway. Elevator rails or other hoistway equipment are not to be used as the down conductor for lightning protection systems.

C. Main feeders for supplying power to elevators are to be installed outside the hoistway unless as follows:

1. By special permission, feeders for elevators are permitted within an existing hoistway if no conductors are spliced within the hoistway.

2. Feeders are permitted inside the hoistway for elevators with driving machine motors located in the hoistway or on the car or counterweight.

620.38, Electrical Equipment in Garages or Similar Occupancies Used for Elevators, states that it is to comply with Article 511.

Part V. Traveling Cables

620.41, Suspension of Traveling Cables. Traveling cables are to be suspended at the car and hoistway end or counterweight end, where applicable, so as to reduce the strain on the individual copper conductors to a minimum.

Traveling cables are to be supported by one of the following means:

1. By their steel supporting member(s)

2. By looping the cables around supports by means that automatically tighten around the cable when tension is increased for unsupported lengths up to 200 feet

Unsupported length for the hoistway suspension means is that length of the cable measured from the point of suspension in the hoistway to the bottom of the loop with the elevator car located at the bottom landing. Unsupported length for the car suspension means is that length of cable measured from the point of suspension on the car to the bottom of the loop, with the elevator car located at the top landing.

620.42. In hazardous (classified) locations, traveling cables are to be a type approved for hazardous locations.

620.43, Location of and Protection for Cables. Traveling cable supports are to be located so as to reduce to a minimum the possibility of damage due to the cables coming in contact with the hoistway construction or equipment in the hoistway. Where necessary, suitable guards are to be provided to protect the cables against damage.

620.44, Installation of Traveling Cables. Traveling cables that are suitably supported and protected from physical damage are permitted to be run without the use of a raceway in either or both of the following:

1. When used inside the hoistway, on the elevator car, hoistway wall, counterweight or controllers and machinery that are located inside the hoistway, provided the cables are inside the original sheath.

2. From inside the hoistway, to elevator controller enclosures and to elevator car and machine room, control room, machinery space and control space connections that are located outside the hoistway for a distance not exceeding six feet in length as measured from the first point of support on the elevator car or hoistway wall or counterweight where applicable, provided the conductors are grouped together and taped or corded or in the original sheath. The traveling cables are permitted to be continued to the equipment.

Part VI. Disconnecting Means and Control

620.51, Disconnecting Means. A single means for disconnecting all ungrounded main power supply conductors for each elevator is to be provided and be designed so that no pole can be operated independently. Where multiple driving machines are connected to a single elevator, there is to be one disconnecting means to disconnect the motor(s) and control the operating magnets.

The disconnecting means for the main power supply conductors is not to disconnect the branch circuit required in 620.22, 620.23 and 620.24.

A. **Type.** The disconnecting means is to be an enclosed, externally operable fused motor switch or circuit breaker that is lockable in accordance with 110.25. The disconnect is to be a listed device.

B. **Operation.** No provision is to be made to open or close this disconnecting means from any other part of the premises. If sprinklers are installed in hoistways, machine rooms, control rooms, machinery spaces or control spaces, the disconnecting means is permitted to automatically open the power supply to the affected elevator(s) prior to the application of water. No provision is to be made to automatically close this disconnecting means. Power is to be restored only by manual means.

C. **Location.** The disconnecting means is to be located where it is readily accessible to qualified persons.

On elevators with generator field control, the disconnecting means is to be located within sight of the motor controller for the driving motor of the motor-generator set. Driving machines, motor-generator sets of motion, and operation controllers not within sight of the disconnecting means are to be provided with a manually operated switch installed in the control circuit to prevent starting. The manually operated switch(s) is to be installed adjacent to this equipment.

Where the driving machine or the motor-generator set is located in a remote machine room or remote machinery space, a single means for disconnecting all ungrounded main power supply conductors is to be provided and be lockable open in accordance with 110.25.

D. Identification and Signs

1. Where there is more than one driving machine in a machine room, the disconnecting means are to be numbered to correspond to the identifying number of the driving machine that they control.

    The disconnecting means is to be provided with a sign to identify the location of the supply side over-current protective device.

2. Available Short-Circuit Current Field Marking. Where an elevator control panel is used, it is to be legibly marked in the field with the maximum available short-circuit current at its line terminals. The field marking(s) is to include the date the short-circuit calculation was performed and be of sufficient durability to withstand the environment involved.

    When modifications in the electrical installation affect the maximum available short-circuit current at the control panel, this figure is to be verified or recalculated as necessary to ensure the elevator's control panel rating is sufficient for the maximum current at the line terminals of the equipment. The required field marking(s) is to be adjusted to reflect the new level of maximum available short-circuit current.

E. Where any of the disconnecting means in 620.51 has been designated as supplying an emergency system load, surge protection is to be provided.

620.52, Power from More than One Source

A. On single-car and multi-car installations, equipment receiving electrical power from more than one source is to be provided with a disconnecting means for each source of electrical power. The disconnecting means is to be in sight of the equipment served.

B. Warning Sign for Multiple Disconnecting Means. Where multiple disconnecting means are used and parts of the controllers remain energized from a source other than the one disconnected, a warning sign is to be mounted on or next to the disconnecting means. The sign is to be clearly legible and is to read as follows:

WARNING

PARTS OF THIS CONTROLLER ARE NOT DE-ENERGIZED
By THIS SWITCH

C. Interconnection Between Multicar Controllers. Where interconnections between controllers are necessary for the operation of the system on multicar installations that remain energized from a source other than the one disconnected, a warning sign is to be mounted on or next to the disconnecting means.

620.53, Car Light, Receptacle(s) and Ventilation Disconnecting Means. Elevators are to have a single means for disconnecting all ungrounded car light receptacle(s) and ventilation power supply conductors for that elevator car.

The disconnecting means are to be an enclosed, externally operable fused motor-circuit switch or circuit breaker that is lockable in accordance with 110.25 and is to be located in the machine room or control room for that elevator car. Where there is no machine room or control room, the disconnecting means is to be located in a machinery space or control space outside the hoistway that is readily accessible to only qualified persons.

Disconnecting means are to be numbered to correspond to the identifying number of the elevator car whose light source they control. The disconnecting means is to be provided with a sign to identify the location of the supply side over-current protective device.

620.54, Hoistway and Air-Conditioning Disconnecting Means. Elevators are to have a single means for disconnecting all ungrounded car heating and air-conditioning power supply conductors for that elevator car.

The disconnecting means is to be an enclosed, externally operable fused motor-circuit switch or circuit breaker that is lockable open and is to be in the machine room or control room for that elevator car. Where there is no machine room or control room, the disconnecting means is to be located in a machinery space or control space outside the hoistway that is readily accessible to only qualified persons.

Where there is equipment for more than one elevator car in the machine room, the disconnecting means is to be numbered to correspond to the identifying number of the elevator car whose heating and air-conditioning source they control.

The disconnecting means is to be provided with a sign to identify the location of the supply side over-current protective device.

620.55, Utilization Equipment Disconnecting Means. Each branch circuit for the utilization equipment is to have a single means for disconnecting all ungrounded conductors. The disconnecting means is to be lockable open. Where there is more than one branch circuit for other utilization equipment, the disconnecting means is to be numbered to correspond to the identifying number of the equipment served. The disconnecting means is to be provided with a sign to identify the location of the supply side over-current protective device.

Part VII. Over-Current Protection Is to Be Provided in Accordance with 620.61 (A) Through (D).

A. Operating devices and control and signaling circuits are to be protected against over current in accordance with the requirements of 725.43 and 725.45. Class 2 power-limited circuits are to be protected against over current in accordance with the requirements of Chapter 9, notes to tables 11(A) and 11(B).

B. Overload Protection for Motors. Motor and branch-circuit overload protection is to conform to Article 430, Part III and (B) (1) through (B) (4).

Duty rating on elevator and motor-generator sets driving motors is to be intermittent. Such motors are permitted to be protected against overload in accordance with 430.33.

C. Motor feeder short-circuit and ground-fault protection is to be as required in Article 430, Part V.

D. Motor branch-circuit, short-circuit, and ground-fault protection is as required in Article 430, Part IV.

620.62, Selective Coordination. Where more than one driving machine disconnecting means is supplied by a single feeder, the over-current protective devices in each disconnecting means are to be selectively coordinated with any other supply side over-current protective devices.

Selective coordination is to be done by a licensed professional engineer or other qualified person engaged primarily in the design, installation, or maintenance of electrical systems. The selection is to be documented and made available to those authorized to design, install, inspect, maintain, and operate the system.

Part VIII. Machine Rooms, Control Rooms, Machinery Spaces and Control Spaces

620.71, Grounding Equipment. Elevator driving machines, motor-generator sets, motor controllers, and disconnecting means are to be installed in a room or space set aside for that purpose unless otherwise permitted in 620.71(A) or (B). The room or space is to be secured against unauthorized access.

A. Motor controllers are permitted outside the spaces specified provided they are in enclosures with doors or removable panels that are capable of being locked in the closed position and the disconnecting means is located adjacent to or is an integral part of the motor controller.

B. Driving Machines. Elevators with driving machines located on the car, on the counterweight, or in the hoistway are permitted outside the spaces herein specified.

Part IX. Grounding

620.81, Metal Raceways, Type MC Cable, Type MI Cable or Type AC Cable attached to elevator cars are to be bonded to metal parts of the car that are bonded to the equipment grounding conductor.

620.82. For electric elevators, the frames of all motors, elevator machines, controllers and the metal enclosures for all electrical equipment in or on the car or in the hoistway are to be bonded in accordance with Article 250, Parts V and VII.

620.83. For elevators other than electric having any electrical conductors attached to the car, the metal frame of the car, where normally accessible to persons, is to be bonded in accordance with Article 250, Parts V and VII.

620.85, Ground-Fault Circuit-Interrupter Protection for Personnel. Each 125-volt, single-phase, 15- and 20-ampere receptacle installed in pits, in hoistways, or on cars of elevators associated with wind turbine elevators is to be of the ground-fault circuit-interrupter type.

All 125-volt, single-phase, 15- and 20-ampere receptacles installed in machine rooms, control spaces, and control rooms are to have ground-fault circuit-interrupter protection for personnel.

A single receptacle supplying a permanently installed sump pump is not required to have ground-fault circuit-interrupter protection.

Part X. Emergency and Standby Power Systems

620.91. An elevator(s) is permitted to be powered by an emergency or standby power system.

A. Regenerative Power. For elevator systems that regenerate power back into a power source that is unable to absorb the regenerative power under overhauling elevator load conditions, a means is to be provided to absorb this power.

B. Other building loads such as power and lighting are permitted as energy absorption means required in 620.91(A) provided that such loads are automatically connected to the emergency or standby power system operating the elevator and are large enough to absorb the elevator regenerative power.

C. The disconnecting means required by 620.51 is to disconnect the elevator from both the emergency or standby power system and the normal power system.

Where an additional power source is connected to the load side of the disconnecting means, which allows automatic movement of the car to permit evacuation of the passengers, the disconnecting means required in 620.51 is to be provided with an auxiliary contact that is positively opened mechanically, and the opening is not to be solely dependent upon springs. This contact is to cause the additional power to be disconnected from its load when the disconnecting means is in the open position.

Article 620 contains a large amount of information pertaining specifically to the electrical design and installation of elevator systems. It makes reference to other NEC articles including Article 110, Requirements for Electrical Installations; 250, Grounding and Bonding; 430, Motors, Motor Circuits and Controllers; and 725, Class 1, Class 2 and Class 3 Remote-Control, Signaling and Power-Limited Circuits. These sections are especially relevant to elevator systems, but actually the electrical work covered therein must comply with the entire NEC, which contains minimum requirements for safe installations. Many jurisdictions permit various types of electrical maintenance to be performed by unlicensed workers under the supervision of licensed individuals or under a corporate license. Everyone involved should ascertain licensing requirements in the particular jurisdiction and, by adhering to NEC requirements, ensure that the installation is fully compliant.

Elevators generally have one or more cars that are capable of carrying persons up and down in vertical hoistways, frequently located in the central portion of a tall building. Most elevators situated in buildings rising over four stories are traction as opposed to hydraulic elevators. The car is raised by means of a set of steel ropes rolling over a deeply-grooved sheave. The weight of the car is balanced by a counterweight sized so that the motor works an equal amount, whether with an average load it is traveling up or down. Elevators can be constructed so that two cars, moving in opposite directions, are counterweights for one another, but there are obvious disadvantages in this arrangement.

Elevators are by far the safest form of mass transit. It is more hazardous to cross the street to get to a building than to ride the elevator car to its top floor. The majority of accidents in elevators, especially fatalities, involve maintenance workers and installers who lose balance at or walk through an open door and fall to the bottom of a darkened hoistway. The few elevator accidents that occur involving post-installation passengers are primarily due to failed door interlocks or other safeguards. Their rarity can be attributed to the high degree of cooperation and professionalism among manufacturers, installers, and maintenance workers.

Modern traction elevators are generally driven by VFD-controlled AC induction motors. These systems can be geared or gearless. In a geared installation, worm gears control the motion of the car. Motor speed and direction respond to signals from the motion controller, which usually sends a pulse-width modulated voltage to the inverter section of the VFD, which in turn conveys a high-voltage, high-current having the same waveform to the motor.

Verification, or in case of faulty operation, diagnosis, can best be performed using a hand-held, battery-powered oscilloscope, which can be set to display the low-voltage signal coming out of the motion controller or the high-voltage feed from the VFD to the motor. Check both ends of this transmission line, because very often the entire fault is in a damaged cable.

Since the motor is high-RPM, much too fast for the elevator lift machinery, speed is reduced and torque increased by means of the gear box. As in an automotive transmission, since there is no combustion, oil changes are not very frequent.

In gearless traction elevators, the motor is connected directly to the sheave with no gearbox. This can be accomplished by having a larger sheave, but instead the usual arrangement involves a low-RPM, higher torque motor. In either type, the brake is located between the motor and the sheave, or sometimes at the outboard end of the sheave. To stop the elevator, the motion controller interrupts power to both the motor and the brake. This also occurs if there is a power outage. The brake is applied only when power is removed. The motor and brake work together. Most brakes are disc brakes with calipers.

Cables are attached to a hitch plate at the top of the car, or they may pass under the car. In the simplest configuration, these cables pass over the top sheave and connect to the counterweight, which is located in the hoistway and has a set of rails that is separate from the car rails.

Depending upon building height and car capacity, there are typically two to twelve ropes, any one of which is capable of supporting a fully-loaded car plus 25 percent additional weight. Rope and sheave weight are greatly reduced by maintaining equal tension on all ropes. The tension should be checked periodically at intervals specified in the operator's manual. Hydraulic rope tensioners are available that maintain equal tension on all ropes. When ropes show excessive wear, they must be replaced. Sheaves can be re-machined. This involves "hanging the cab."

If the elevator has a vertical working distance greater than 98 feet, a separate chain or set of cables is attached to the bottom of the counterweight and to the bottom of the car. This system is known as compensation, and its purpose is to compensate for the increasing weight of steel rope as the car descends. If the compensation means consists of cable, an additional sheave is located in the pit to guide the cable. If it consists of chain, a barrier is necessary to prevent damage to adjacent elevator components.

STUDy QUESTIONS

1. The National Electrical Code:
   A. begins with fundamentals for untrained individuals
   B. is primarily concerned with safety issues
   C. is primarily concerned with energy efficiency
   D. is an excellent guide for reducing costs in electrical installations

2. Specific requirements for elevator systems are in NEC Article:
   A. 600
   B. 610
   C. 620
   D. 630

3. The control room:
   A. is outside the hoistway
   B. can be in the pit
   C. contains the driving machine or hydraulic machine
   D. may not contain electrical and/or mechanical equipment used directly in connection with the elevator

4. The machine room:
   A. contains the electrical driving machine
   B. may be used for general storage
   C. is not required to be locked
   D. must not have electrical receptacles

5. Operating devices have _____ that activate the operation controller.
   A. car switches
   B. push buttons
   C. key or toggle switches
   D. any of the above

6. Voltage Limitations provides that the supply voltage is not to exceed _____ volts between conductors unless permitted.
   A. 250
   B. 300
   C. 350
   D. 600

7. Power Circuits provides that branch circuits to driving machine motors are not to have a voltage in excess of _____ volts.
   A. 250
   B. 300
   C. 600
   D. 1000

8. Conductors and optical fibers located in hoistways, machinery spaces, control spaces, in or on cars, in machine rooms and control rooms, not including traveling cables, are to be:
   A. in conduit, EMT or wireways
   B. Type MC cable
   C. Type MI or AC cable
   D. any of the above

9. The following are permitted in hoistways in lengths not to exceed six feet:
   A. flexible metal conduit
   B. liquidtight metal conduit
   C. liquidtight flexible metallic conduit
   D. any of the above

10. The following are permitted on the car assembly in lengths not to exceed six feet:
    A. flexible metal conduit
    B. liquidtight flexible metal conduit
    C. liquidtight flexible nonmetallic conduit
    D. any of the above

For answers, go to Appendix A.

CHAPTER 9

MOTION CONTROLLER

Motion controller is a term that is used in many fields—robotics, avionics, and computer-controlled electromechanical machinery of all types. It implies X, Y and Z axes, but in elevators, the motion is mostly vertical, plus doors that open and close, and rotary motion of the electric motor. Always, the motion is controlled in a highly precise manner.

To begin, we will consider the generic motion controller, as applicable not only to elevator systems, but to any electrical system that involves energy and motion. The most basic motion controller is a motor controller.

NEC in Article 430 states that all electrical motors are to have suitable controllers. For stationary motors of 1/8 horsepower or less that are left running and cannot be damaged by overload or failure to start, they can be as simple as a branch-circuit disconnecting means. One example is a wall clock that is protected by a circuit breaker or fuse in an entrance panel or load center, as shown in Figure 9-1. A further level of protection is required for a portable motor that is rated 1/3 horsepower or less. The controller for this motor is permitted to be an attachment plug and receptacle or cord connector.

Branch-circuit inverse-time circuit breakers rated in amps are permitted controllers for all motors. They may also serve as overload protection. For stationary motors rated at less than two horsepower and 300 volts, controllers may be general-use switches having a rating of not less than twice the full-load current of the motors. On AC circuits, general-use snap switches suitable for use only on AC (not general-use AC-DC snap switches) are permitted as controllers, where the motor full-load current rating is not more than 80 percent of the ampere rating of the switch.

If the controller also serves as the disconnecting means for the motor, the controller is not required to open all conductors to the motor. For a 240-volt, single-phase motor, unlike an in-circuit switch, the controller could open one leg only, and this would stop the motor, though one of the terminals would remain live. In a three-phase motor, two legs would have to be opened to ensure the motor would not con-

FIGURE 9-1　A circuit breaker is permitted to serve as a motor controller.

tinue to run on two phases. With exceptions, each motor is to be provided with an individual controller.

The purpose of a motion controller, shown in Figure 9-2, is to regulate the operation of moving parts by means of downstream actuators. Human intervention is not required. A motor, linear or rotary, is one type of machine. Others machines   include devices that produce heat, ventilation, chemical processes, locomotion, earth-moving and aeronautic equipment, and many types of instrumentation. Beyond the controller may be an energy amplifier, memory, and display.

Motion controllers may be open-loop or closed-loop, the latter made possible by feedback circuitry. This usually allows the motion controller to sense and act upon the position of the actuator.

An open-loop system often consists of a stepper motor, which sets the position or rotary velocity of its load. A servomotor does the same thing but additionally it conveys information to the controller regarding the status of the connected load, permitting error correction, which can occur continuously or take place intermittently. This is not needed for an air fan or pump moving liquid, although usually there is a sensor in the reservoir indicating liquid level, or a low On sensor and a high Off sensor providing the hysteresis. Closed-loop functionality can be conveyed through a power cable, through a separate data cable, or wirelessly. Closed-loop technology often uses incremental encoders or Hall effect sensors to convey velocity, position, and torque information to the controllers, which then perform real-time error correction.

FIGURE 9-2 Motion controller for an elevator.

Besides single- or multi-axis regulation of the actuator, a controller can create a response profile, allowing a continuous range of position or velocity. The motion controller then generates torque commands, which are received by the drive and which fashion the motor response in terms of position and velocity. Motion controllers implement motor response by means of proportional control, in which the gain is constant. It may be derivative or integral. These constitute PID, the dominant control algorithm at this time.

The principle components in PID are setpoint and feedback. Setpoint is set by the programmer. Feedback derives from position, velocity, temperature, or other sensors in the driven load. Setpoint is an ideal state but feedback is based on the actual state. The difference between these two parameters is error. The controller always attempts to reduce error to zero.

Error changes result from fluctuations in the setpoint or in the load. The controller then increases or decreases the applied torque, adjusts vector, or generally does what it takes to align feedback and setpoint. PID uses proportional, integral, and derivative concepts. In a static system, where setpoint and feedback are constant, the controller maintains uniform torque. But most systems are dynamic, so torque has to be adjusted, either occasionally or at millisecond intervals.

Integration is the sum of a function in time. In PID control, integration and derivation, which is the rate of change, are operative. When the controller is working as intended, P, I and D interact to minimize error. If there is no error or change in setpoint, P does not change.

Proportionality is the product of gain and error. The controller continually adjusts it to eliminate error. Setting it too high causes the controller to exceed the setpoint, leading to oscillation.

Regardless of magnitude, eventually error increases, and that is where the integral factor becomes operative. Now, control loop action is required, and it adjusts the P factor to reduce error without initiating oscillation.

The derivative factor is not always needed to correct present error relative to previous error. Instead it is more concerned with the rate of change of this error. An increase in error rate of change or a longer time period results in a larger derivative error factor. The controller responds by accelerating its error change rate, suppressing any tendency of P and I to depart from the set point. The derivative factor reduces P and I in proportion to the derivative time.

The PID controller quantifies the current error integrated in time and the derivative or rate of change of the error and applies the correction to modify the feedback. The error calculation update rate varies widely, sometimes occurring thousands of times per second.

Actuator motion can be applied by means of mechanical components to realize the desired motion. Equipment is usually assembled to include gears, shafts, screws, belts, linkages, and linear and rotational bearings.

Besides profiling, single-axis control has evolved to permit numerous axes of motion. The first interfaces between the motion controller and driven machinery were analog, but as greater precision and speed have evolved, digital control and feedback have taken over. Interfaces that unite drive and driven machinery are increasingly automatic and instantaneous.

Programmable Logic Controllers

Programmable logic controllers (PLCs), as shown in Figure 9-3, are used everywhere. One reason for their success is the great simplicity of their elegant form of programming by means of highly-intuitive ladder diagrams. Since first implemented by auto manufacturers in the 1960s, factory electricians have been able to set inputs and outputs for new machinery as it is powered up.

This programming language is graphical rather than text-based. What is unique about PLC programming is that it is comprised of electrical diagrams with virtual switches, relays, and loads connected between two control lines like rungs in a ladder. The contacts, coils, and other devices are not real components. They refer to elements within the PLC memory. Technicians draw ladder diagrams on a PC screen and download them into the PLC, which is cabled to sensors and actuators on the assembly line. Proprietary software, of course, is required and it is supplied by the PLC manufacturer. The circuitry enables the PLC to work in conjunction with VFD to control speed, direction, and torque of a conventional AC induction motor. PLCs in this manner can be quickly configured to control any type of electrical machinery without resorting to hardwire techniques.

FIGURE 9-3 Allen-Bradley PLC.

The American Society of Mechanical Engineers (ASME), through its Board of Safety Codes and Standards, develops and maintains a comprehensive portfolio of mandates that govern existing elevators and escalators. A complete listing can be seen at ASME.org. ASME A17, Safety Code for Existing Elevators and Escalators, is an essential reference for elevator maintenance workers, repair technicians, and manufacturers. It can be ordered online at the website.

Because of its central position in elevator technology, we reiterate some of the provisions:

Section 3.8.1, General Requirements, provides that sheaves and drums are to be made of cast iron or steel with finished grooves for ropes. (In referring to ropes, of course, steel ropes are meant.) Set screw fastenings are not to be used in lieu of keys or pins on connections subject to torque or tension. Friction gearing or a clutch mechanism is not to be used to connect a driving mechanism, other than in connection with a car leveling device.

Section 3.8.2, Winding Drum Machines, provides that these machines are to be provided with a slack-rope device having an enclosed switch of the manually reset type that will cause the electric power to be removed from the elevator driving machine motor and brake if the hoisting ropes become slack or broken.

Final terminal stopping devices for winding drum machines are to consist of a stopping switch located on the driving machine and a stopping switch located in the hoistway and operated by cams attached to the car.

Stopping switches, located on and operated by the driving machine, are not to be driven by chains, ropes, or belts. The opening of these contacts is to occur before or coincident with the opening of the final terminal stopping switch.

Where a three-phase AC driving-machine motor is used, the mainline circuit to the driving-machine brake coils are to be directly opened either by the contacts of the machine stop switch or by stopping switches mounted in the hoistway and operated by a cam attached to the car.

Driving machines equipped with a DC motor and DC brake are permitted to have the final terminal stopping device contacts installed in the operating circuits. The occurrence of a single ground or failure of any single magnetically operated switch, contactor or relay must not render any final terminal stopping device ineffective.

Section 3.8.4, Brakes, provides that the elevator driving machine is to be equipped with a friction brake applied by a spring or springs, or by gravity, and released electrically. The brake is to have a capacity sufficient to hold the car at rest with its rated load. For passenger elevators and freight elevators permitted to carry employees, the brake is to be designed to hold the car at rest with an additional load up to 25 percent in excess of the rated load.

Section 3.9, Terminal Stopping Devices, states that enclosed upper and lower normal stopping devices are to be provided and arranged to slow down and stop the car automatically at or near the top and bottom terminal landings. Such devices are to function independently of the operation of the normal stopping means and of the final terminal stopping device.

Normal stopping devices are to be located on the car, in the hoistway or in the machine room, and are to be operated by the movement of the car.

Broken rope, tape or chain switches are to be provided in connection with normal stopping devices located in the machine room of traction elevators. Such switches are to be opened by a failure of the rope, tape or chain and are to cause the electrical power to be removed from the driving-machine motor and brake.

Section 3.9.2, Final Terminal Stopping Devices, states that enclosed upper and lower final terminal electromechanical stopping devices are to be provided and arranged to prevent movement of the car by the normal operating devices in either direction of travel after the car has passed a terminal landing. Final terminal stopping devices are to be located as follows:

Elevators having winding drum machines are to have stopping switches on the machines and also in the hoistway operated by the movement of the car.

Elevators having traction-driving machines are to have stopping switches in the hoistway operated by the movement of the car.

Section 3.10.4, Electrical Protective Devices, states that these devices are to be protected as follows:

    a. **Slack-Rope Switch.** Winding drum machines are to be provided with a slack-rope device equipped with a slack rope switch of the enclosed manually reset type that will cause the electric power to be removed from the elevator driving-machine motor and brake if the suspension ropes become slack.

b. **Motor-Generator Running Switch.** Where generator-field control is used, means are to be provided to prevent the application of power to the elevator driving machine motor and brake unless the motor-generator set connections are properly switched for the running condition of the elevator. It is not required that the electrical connections between the elevator driving-machine motor and the generator be opened in order to remove power from the elevator motor.

c. **Compensating Rope Sheave Switch.** Compensating rope sheaves are to be provided with a compensating rope sheave switch or switches mechanically opened by the compensating rope sheave before the sheave reaches its upper or lower limit of travel to cause the electric power to be removed from the elevator driving-machine motor and brake.

d. **Broken Rope, Tape or Chain Switches Used in Connection with Machine Room Normal Stopping Switches.** Broken rope, tape or chain switches are to be provided in connection with normal terminal stopping devices located in machine rooms of traction elevators. Such switches are to be opened by failure of the rope, tape or chain.

e. **Stop Switch on Top of Car.** A stop switch is to be provided on the top of every elevator car, which will cause the electric power to be removed from the elevator driving-machine motor and brake. It is to be of the manually operated and closed type with red operating handles or buttons. It is to be conspicuously and permanently marked "STOP" and indicate the stop and run positions. It is to be positively opened mechanically, opening not solely dependent on springs.

f. **Car Safety Mechanism Switch.** A switch is required where a car safety is provided.

g. **Speed Governor Overspeed Switch.** Where required by Section 3.6.1, a speed governor overspeed switch is to be provided.

h. **Final Terminal Stopping Device.** Where reduced stroke oil buffers are provided, final terminal stopping devices are to be provided.

i. **Emergency Terminal Speed Limiting Device.** Where reduced stroke oil buffers are provided, emergency terminal speed limiting devices are required.

j. **Motor Generator Overspeed Protection.** Means are to be provided to cause the electric power to be removed automatically from the elevator driving-machine motor and brake, should a motor generator overspeed excessively.

k. Motor Field Sensing Means. Where DC is supplied to an armature and shunt field of an elevator driving-machine motor, a motor-field current sensing means is to be provided, which will cause the electric power to be removed from the motor armature and brake unless current is flowing in the shunt field of the motor. A motor-field current sensing means is not required for static control elevators provided with a device to detect an overspeed condition prior to, and independent of, the operation of the governor overspeed switch. This device must cause power to be removed from the elevator driving-machine motor armature and machine brake.

l. Buffer Switches for Oil Buffers Used with Type C Car Safeties. Oil level and compression switches are to be provided for all oil buffers used with Type C safeties.

m. Hoistway Door Interlocks or Hoistway Door Electric Contacts. Hoistway door interlocks or hoistway door electric contacts are to be provided for all elevators.

n. Car Door or Gate Electric Contacts. Car door or gate electric contacts are to be provided for all elevators.

o. Normal Terminal Stopping Devices. Normal terminal stopping devices are to be provided for every elevator.

p. Car Side Emergency Exit Electric Contact. An electric contact is to be provided on every car side emergency exit door.

q. Electric Contacts for Hinged Car Platform Sills. Where provided, hinged car platform sills are to be provided with electric contacts.

r. In-Car Stop Switch. On passenger elevators equipped with nonperforated enclosures, a stop switch, either key-operated or behind a locked cover, is permitted to be provided in the car and located in or adjacent to the car operating panel. The switch is to be clearly and permanently marked "STOP" and is to indicate the stop and run positions. The switch is to be positively opened mechanically and its opening is not to be solely dependent on springs. When opened, this switch is to cause the electric power to be removed from the elevator driving-machine motor and brake.

s. Emergency Stop Switch. On all freight elevators, passenger elevators with perforated enclosures and passenger elevators with nonperforated enclosures not provided with an in-car stop switch, an emergency stop switch is to be provided in the car and located in or adjacent to each car operating panel. When open ("STOP" position), this switch is to cause the electric power to be removed from the elevator driving-machine motor and brake and is to:

1. be of the manually operated and closed type

2. have red operating handles or buttons

3. be conspicuously and permanently marked "STOP" and is to indicate the stop and run positions

4. have contacts that are positively opened mechanically (opening not solely dependent on springs)

t. **Stop Switch in Pit.** A stop switch conforming to (e) Stop Switch on Top of Car is to be provided in the pit of every elevator. The switch is to be adjacent to every pit access.

u. **Buffer Switches for Gas Spring Return Oil Buffers.** A buffer switch is to be provided for gas spring return oil buffers that will cause electric power to be removed from the elevator driving-machine motor and brake if the plunger is not within 0.5 inches of the fully extended position.

Section 3.10.5, Power Supply Disconnecting Means provides that:

a. A disconnect switch or a circuit breaker is to be installed and connected into the power supply line to each elevator motor or motor generator set and controller. The power supply line is to be provided with over-current protection, preferably inside the machine room.

b. The disconnect switch or circuit breaker is to be of the manually closed multipole type, and is to be visible from the elevator driving machine or motor generator set. When the disconnecting means is not within sight of the driving machine, the control panel, or the motor-generator set, an additional manually operated switch is to be installed adjacent to the remote equipment and connected to the control circuit to prevent starting.

c. No provision is to be made to close the disconnect switch from any other part of the building.

d. Where there is more than one driving machine in a machine room, disconnect switches or circuit breakers are to be numbered to correspond to the number of the driving machine that they control.

Section 3.10.6, Phase Reversal and Failure Protection, states that elevators having polyphase AC power supplies are to be provided with means to prevent the starting of the elevator motor if the phase rotation is in the wrong direction or if there is a failure of any phase. This protection is considered to be provided in the case of generator field control having AC motor-generator driving motors, provided a reversal of phase will not cause the elevator driving machine motor to operate in the wrong direction. Controllers on which switches are operated by polyphase torque motors provide inherent protection against phase reversal or failure.

Section 3.10.7, Operating of the Driving Machine with a Hoistway Door Unlocked or a Hoistway Door or a Car Door Not in the Closed Position. This is permitted under the following conditions:

a. by a car-leveling or truck zoning device

b. when a hoistway access switch is operated

c. when the top-of-car or in-car inspection operation utilizing a car-door bypass or hoisting-door bypass switch is activated.

Devices other than those specified above are not to be provided to render hoistway-door interlocks, the electric contacts of hoistway-door mechanical locks and electric contacts, or car door, gate electric contacts, or car door or gate interlocks inoperative. Existing devices that do not conform to the above are to be removed.

Section 3.10.8, Release and Application of Driving Machine Brakes, states that driving-machine brakes are not to be electrically released until power has been applied to the driving-machine motor.

Two devices are to be provided to remove power independently from the brake. If the brake circuit is ungrounded, all power lines to the brake are to be opened. The brake is to apply automatically when:

a. The operating device of a car switch or continuous pressure operation switch is in the stop position

b. A floor stop device functions

c. Any of the electrical protective devices in Section 3.10.4 functions

Under conditions described in (a) or (b), the application of the brake is permitted to occur on or before the completion of the slowdown and leveling operations.

The brake is not to be permanently connected across the armature or field of a DC elevator driving motor.

Section 3.10.9, Control and Operating Circuit Requirements, provides that the failure of any magnetically operated switch, contactor or relay to release in the intended manner, or the occurrence of a single accidental ground, or combination of accidental grounds, is not to permit the car to start or run if any hoistway door interlock is unlocked or if any hoistway door or car door or gate electric contact is not in the closed position.

Section 3.10.10, Absorption of Regenerated Power, states that when a power source is used which, in itself, is incapable of absorbing the energy generated by an overhauling load, means for absorbing sufficient energy to prevent the elevator from attaining governor tripping speed or a speed in excess of 125 percent of rated speed, whichever is lesser, is to be provided on the load side of each elevator power supply line disconnecting means.

Section 3.10.12, System to Monitor and Prevent Automatic Operation of the Elevator with Faulty Door Contact Circuits, states that means are to be provided to monitor the position of power-operated car doors that are mechanically coupled with the landing doors while the car is in the landing zone in order:

a. to prevent automatic operation of the car if the car door is not closed, regardless of whether the portion of the circuits incorporating the car door contact or interlock contact of the landing door coupled with the car door, or both, are closed or open, except as permitted in 3.10.7

b. to prevent the power closing of the doors during automatic operation if the car door is fully open and if any of the following conditions exist:

1. The car door contact is closed, or a portion of the circuit incorporating this contact is bypassed

2. The interlock of the landing door that is coupled to the opened car door is closed, or the portion of this circuit incorporating this contact is bypassed

3. The car door contact and the interlock contact of the door that is coupled to the opened car door are closed, or the portions of the circuits incorporating these contacts are bypassed

Section 3.11, Emergency Operation and Signaling Devices, Section 3.11.1 Car Emergency Signaling Devices, states that in all buildings, the elevator(s) are to be provided with the following:

a. If installed, altered or both under ASME A17.1 – 2000 or earlier edition

1. an audible signaling device, operable from the emergency stop switch, when provided, and from a switch marked "alarm" that is located in or adjacent to each car operating panel. The signaling device is to be located inside the building and audible inside the car and outside the hoistway. One signaling device is permitted to be used for a group of elevators.

2. means of two-way communication (telephone, intercom, etc.) between the car and a readily accessible point outside the hoistway that is available to emergency personnel. The means to activate the two-way communication system does not have to be provided in the car.

3. If the audible signaling device, or the means of two-way communication, or both, are normally connected to the building power supply, they are to automatically transfer to a source of emergency power within ten seconds after the normal power supply fails. The power source is to be capable of providing for the operation of the audible signaling device for at least one hour, and the means of two-way communication for at least four hours.

4. In buildings in which a building attendant (building employee, watchman, etc.) is not continuously available to take action when the required emergency signal is operated, the elevators are to be provided with a means within the car for communicating with or signaling to a service that is capable of taking appropriate action when a building attendant is not available.

5. An emergency power system is to be provided conforming to the requirements of (a) (3).

b. If installed, altered or both under ASME A17.1a—2002 or later editions, the emergency communications system is to comply with Section 2.27 of the ASME A17.1/CSA B44 Code under which it was installed or altered.

Section 3.11.2, Operations of Elevators Under Standby (Emergency) Power, states that an elevator is permitted to be powered by a standby (emergency) power system provided that, when operating on such standby power, there is conformance to the requirements of Section 3.10.10.

Section 3.11.3, Firefighters' Service provides that elevators are to conform to the requirements of ASME/ANSI A17.1—1987 Rules 211.3 through 211.8 unless at the time of installation or alteration it was required to comply with a later edition of A17.1.

All elevators that are part of a group are to conform to identical firefighters' service operation requirements regardless of which edition of A17.1 they complied with at the time of their installation or alteration.

The Phase I and Phase II switches for all elevators in a building are to be operable by the same key.

Section 3.12, Suspension Means and their Connections, provides that cars are to be suspended by steel wire ropes attached to the car frame or passing around sheaves attached to the car frame. Only iron (low-carbon steel) or steel wire ropes, having the commercial classification "Elevator Wire Rope" or wire rope specifically constructed for elevator use is to be used for the suspension of elevator cars and for the suspension of counterweights.

Section 3.12.4, Minimum Number and Diameter of Suspension Ropes, provides that all elevators except freight elevators that do not carry passengers or freight handlers and have no means of operation in the car are to conform to the following requirements:

a. The minimum number of hoisting ropes used is to be three for traction elevators and two for drum-type elevators. Where a car counterweight is used, the number of counterweight ropes used shall be not less than two.

b. The minimum diameter of hoisting and counterweight ropes is to be 0.375 inches. Outer wires of the ropes are to be not less than 0.024 inches in diameter.

Section 3.12.5, Suspension Rope Equalizers, states that suspension rope equalizers, where provided, are to be of the individual compression spring type.

Section 3.12.6, Securing of Suspension Ropes to Winding Drums, provides that suspension wire ropes of winding drum machines are to have the drum ends of the ropes secured on the inside of the drum by clamps or by tapered babbitted sockets, or by other approved means.

Section 3.12.7, Spare Rope Turns on Winding Drums, provides that suspension wire ropes of winding drum machines are to have not less than one turn of the rope on the drum when the car is resting on the fully compressed buffers.

Section 3.12.8, Suspension Rope Fastenings, states that spliced eyes by return loop are permitted to continue in service. Suspension rope fastenings are to conform to Requirement 2.20.9 of ASME A17.1—2004 when the ropes are replaced.

Section 3.12.9, Auxiliary Rope Fastening Devices, states that auxiliary rope fastening devices, designed to support elevator cars or counterweights if any regular rope fastening fails, are permitted to be provided.

Besides providing a guide for elevator manufacturers in complying with current code requirements, the ASME regulations are useful for technicians in seeing how sensors on the car, throughout the hoistway, and in the machine room report the status of the many safety mechanisms. Typically, when one of these is not able to function, power to the motor and brake is removed, causing the car to stop and remain in place until the problem is rectified. This interruption of service is frequently caused by a sensor malfunction, or a cabling problem between the sensor and the motion controller. By interpreting the error code, the technician can proceed from the motion controller to the sensor and beyond, to isolate and repair the fault.

The motion controller is the nerve center of an elevator system. It has well-defined inputs and outputs that extend throughout the system. Electrical lines, either small copper conductors or complex serial bus circuits, connect to sensors, buttons, and actuators. At each floor there are call buttons, lighted displays that enable passengers to follow the car's progress, gongs to announce its arrival, linear actuators to open and close hoistway and car doors, etc.

Circuits pertaining to the car interior and exterior are conveyed through finely-stranded wires within the traveling cable. Some circuits terminate above the car top, permitting technicians to stop and start car motion and put it in slow-speed inspection mode.

Motion controller lines also terminate at the variable frequency drive and the motor. The elevator brake, in a traction elevator, is located at the motor output. Motor and brake always work together. When current to the motor is interrupted, due either to a power outage or to a command from the motion controller, the motor is powered down. Simultaneously, current to the brake is interrupted and springs cause it to be applied. Otherwise, the motor would coast and the car would continue to move. (To emphasize: Power on = motor runs, brake not applied. Power off = motor stops running, brake applied.)

Additionally, the motion controller varies the motor's speed, by means of the variable frequency drive. At each landing, the motor and car slow down before stopping, providing a comfortable transition for the passengers and reducing mechanical and electrical stress in the elevator system.

The motion controller is usually located in the machine room or in a control space. In machine room-less elevator designs, it may be located in a hallway, usually on the top floor adjacent to the elevator door.

Since it is not kept in a locked machine room, the motion controller enclosure must be locked, with access restricted to authorized persons. Depending on the size and complexity of the installation, the motion controller may be wall-mounted or contained in one or more floor-to-ceiling steel cabinets.

We have emphasized that elevator technicians must avoid going beyond their level of expertise. In working on an elevator motion controller, there is the potential for inadvertently introducing hazards into the system, creating a situation in which passengers can be injured. Door interlock mechanisms in the motion controller prevent the car from moving when a door is open and prevent the door from opening when the car is moving. One of the functions of the motion controller is to prevent a passenger from being caught in a door opening while the car is moving.

In the past, workers in servicing a motion controller have unwittingly disabled door interlocks or similar safety mechanisms, setting the stage for a catastrophic event.

You should not work on elevator controller circuitry unless and until you are fully trained and certified to do this type of work. Even seemingly passive and harmless operations such as taking resistance and continuity readings can have unintended consequences. Today's ubiquitous metal-oxide semiconductor field effect transistors (MOSFETs) and complementary metal-oxide semiconductor (CMOS) devices are incredibly sensitive to static charge and can be destroyed with no visual sign merely by touching them. To avoid this, electronic technicians wear a grounding bracelet when handling a circuit board that may contain CMOS devices.

Elevator motion controllers often contain rows of screw terminals where short insulated copper jumpers are inserted. As part of a maintenance procedure, technicians "jump out" specified pairs of terminals. Don't do this until you have become fully aware of the rationale for the procedure and all of its implications. Mis-jumped terminals can disable a door interlock!

The elevator motion controller is tasked with coordinating total elevator system functionality so that actuators, motor, and brake operate at the right time and car travel, speed, door opening and closing, car leveling, and in-car and lobby indicators operate in the correct sequence. All of this is in response to sensors located throughout the hoistway, in the machine room, and on the car. To this end, inputs such as button signals result in predictable outputs such as in lines to the motor and brake.

The most common elevator controller types are:

- Selective operation, which remembers and answers calls in one direction and then reverses. When the trip has completed, the car returns to the ground floor.

- Group automatic operation, in which two or more elevators are controlled by means of microprocessors to prioritize and respond to calls in an efficient and coordinated manner so as to provide rapid service without excessive, redundant stops.

The control system is comprised of inputs, outputs, and controller. The inputs originate at sensors, button, key controls such as fire service, and system controls. The sensors are magnetic, photo-electric, infrared, weight sensors, and velocity transducers.

The magnetic and photo-electric sensors convey information to the motion controller regarding the car location. The sensor is located on the car and tracks the car location by sensing the number of holes in a guide rail. The magnetic sensor gathers this information by counting magnetic pulses. The infrared sensor is located at the car door and responds to passengers entering or exiting the car. The weight sensor responds to overload and activates an alarm in the car.

The velocity transducer responds to an encoder on the drive sheave. Besides its normal speed, the car slows down as it approaches a landing before stopping, in order to provide a more comfortable ride. An elevator technician can cause the car to creep along at reduced speed by placing it in inspection mode to facilitate safe servicing. This is accomplished by operating a switch on the car top, in the pit, or inside the motion controller enclosure.

There are a number of buttons, all wired back to the motion controller. Most of these buttons are illuminated, lighting up from within to indicate that they are functioning when pressed. Buttons include hall buttons to call a car, floor request buttons inside the car, an open door button, stop button, and ring button to request assistance. Indicators, inside the car and at each floor, indicate the car location.

Key controls can be operated only by inserting and turning a key. They are located in the car and at the ground floor. Included among them are firefighters' service, Phase I and Phase II and an inspector's switch, which places the elevator in low-speed inspection mode for the purpose of maintenance and repair. In this mode, there are manual up and down controls, to be operated only by the elevator technicians. System controls are located in the machine room or elevator control room. These controls can be used to turn the elevator system off when the building is not in use.

Outputs, wired from the motion controller, include actuators, bells, and displays. Among the actuators are the electric motor and brake and the door opening and closing device. Older elevators had a single-phase, fractional horsepower motor located on the car top, with a V-belt and pulley arrangement for operating the doors. Newer designs consist of linear actuators that use linkages to operate the car and hoistway doors.

The emergency bell, operated by a button in the car, indicates that the car is not moving and one or more passengers is trapped. A load bell inside the car indicates that excess weight has been brought into the car and that the elevator cannot be operated safely.

The principle types of motion controllers are:

- **Relay-based.** This is an older technology, still seen in some elevator systems, particularly where there are few stops and manual door operation. It is characterized by switches that are opened and closed by electromagnets. Servicing can be difficult due to the number of relays, limited working space, and labor-intensive solder joints.
- **Solid-state logic technology.** This is characterized by greater reliability, lower power consumption, and technician-friendly fault diagnostics.
- **PLC-based technology.** This is based on a robust, very reliable industrial computer that is capable of controlling very complex systems. Considering the size, the systems are relatively easy to diagnose and repair.

The elevator motion controller is located in an appropriately-sized enclosure, most typically in the machine room. There are some remotely located components, with discrete power supplies, but connected electrically to the motion control module.

The car operating panel, in an automatic elevator, is used primarily by passengers in the car. At eye level, generally located adjacent to the car door, is the panel, which fits into a cutout in the interior wall. It has an oversize metal flange that covers the cutout so as to provide a finished appearance.

This panel in turn also has a cutout, in which the housing is mounted. It also has a cover, which extends beyond the cutout. In the housing is mounted the control buttons and indicators. Major elevator manufacturers such as Otis offer a range of car models, some with plain metal interiors and others with elaborate hardwood interiors, upholstered benches, and upscale lighting. In all cases, the appearance of the control panel is appropriate for and blends in with the car décor. Electrically, all car control panels are similar. In the elevator button panel, there are numbered floor request buttons for each of the destinations. These are typically in two vertical rows with buttons for odd-numbered floors on the left and for even-numbered floors on the right. Below the floor call buttons are the open door button, stop request button, key-operated mode switch, and ring bell button. Additionally, there is a fan button, overload indicator, car position and direction indicator, firefighters' button, and an instruction plate. To diagnose and service wire terminations, it is a simple matter to remove the cover and pull out the car operating panel. Also in the car is a communications interface. This consists of a telephone by which passengers in a stuck elevator can speak to a technician.

Some elevator systems have a supervisory control panel that may be located in the maintenance shop or office. It displays the status of all elevators including car locations and direction of travel.

Specific Operating Modes

Anti-Crime Protection is an operating mode in which each car stops at a specified landing and the doors open so that a security guard can check the occupants.

In Up Peak Mode, elevator cars are recalled to the lobby to serve large numbers of passengers entering the building, generally in the morning as workers arrive or at the conclusion of lunch period. Up Peak may be initiated by a time clock, by simultaneous departure of many cars from the lobby, or by a manual switch.

Down Peak is similar, but it is initiated when many passengers are leaving the building, for example at the end of the day.

Sabbath Service is activated in areas where there are many observant Orthodox Jews, who do not operate electrical devices on the Sabbath. The elevator stops at every floor so that control buttons do not have to be pressed.

Independent service is activated by a key switch in the car or lobby. It disables calls from separate floors. The individual elevator remains parked at a floor with its doors open until a floor is selected. This is useful when moving large or numerous objects to a specific destination.

Medical Emergency (Code Blue Service) is used in health care facilities. It permits a car to be summoned to any floor for use in an emergency. Each floor has a Code Blue switch which, when activated, will call the car that can respond fastest, cancelling scheduled stops. Passengers are notified by an alarm to exit when the car stops.

Car Top Operating Station

Above the car is the car top operating station. Passengers do not have access to this area, and it is used only by an elevator technician during initial installation and servicing of the elevator system. From this location, the car can be put in low-speed inspection mode and be moved slowly up and down in the hoistway. Door opening and closing mechanisms are accessed for repair from this location.

The motion controller enclosure may be in the machine room or for machine room-less designs, wall-mounted in the hallway adjacent to the hoistway door on the top floor. This enclosure may house a relay-based controller, solid state logic controller, or PLC-type controller. The variable frequency drive that operates the elevator motor may be within this enclosure or external to it.

STUDY QUESTIONS

1. A motion controller can be:
   A. a circuit breaker
   B. a plug and receptacle or plug and cord connector
   C. an AC only snap switch
   D. any of the above

2. A motion controller:
   A. operates downstream actuators
   B. is always operated by a person
   C. must be closed-loop
   D. always consists of a stepper motor

3. PLCs:
   A. require human operators
   B. are programmed by means of ladder diagrams
   C. usually are powered by a 480-volt supply
   D. are falling out of favor

4. Ladder programming is:
   A. text-based
   B. graphical
   C. too difficult for maintenance electricians
   D. not used in the latest PLCs

5. When a sensor shows that a safety device is not functioning:
   A. power is interrupted to the motor so that it cannot move the car
   B. power is interrupted to the brake, so that it is applied
   C. the motion controller generates an error message
   D. all of the above

6. Motion controller inputs and outputs:
   A. are often ignored
   B. consist of call button, door interlocks, and numerous sensors and actuators
   C. are not used in hydraulic elevators
   D. are not used in traction elevator

7. The traveling cable consists of:
  A. solid wires
  B. fine stranded wires
  C. armored cable
  D. bare wire without insulation

8. The motion controller:
  A. is not connected to the VFD
  B. is powered by a DC motor
  C. usually has an alphanumeric display
  D. is powered down when the car stops at a landing

9. The motion controller:
  A. is usually in the machine room
  B. is omitted in machine room-less designs
  C. requires a dedicated 600-volt supply
  D. can be easily rebuilt as needed

10. MOSFETs are:
  A. not sensitive to static discharge
  B. being replaced by simple diodes
  C. very common in printed circuit boards
  D. usually the plug-in type

For answers, go to Appendix A.

CHAPTER 10

SYSTEMS CONNECTED TO ELEVATOR INSTALLATIONS AND HOW THEY WORK IN CONCERT

An elevator system, whether it consists of a single car or it is a complex group installation, is closely integrated into the building in which it is situated. It relates to the building's structure (think of the hoistway), electrical system, communication networks, and above all, emergency services. Specific elements, all connected to the central fire alarm control panel, shown in Figure 10-1, include:

- Fire alarm system
- Sprinklers
- Outside telephone system
- Emergency electrical system

Beginning with the first item, we will describe a typical fire alarm system and go on to discuss how it is connected to the elevator system and governs its operation in the event of fire. A centrally controlled and automatically monitored fire alarm system is emphatically unlike the smoke detectors, shown in Figure 10-2, that are highly recommended in residential settings, even if the individual detectors are wired together to sound the alarm in concert.

Residential-type smoke detectors contain individual batteries, usually 9-volt dry cells, and they may or may not be wired into an electrical system branch circuit, which provides redundant power if the battery becomes depleted, while the battery provides power if the electrical supply is interrupted, which is often the case in a fire.

Fire detector heads that are part of a centrally-monitored and controlled fire alarm system, in contrast, do not contain internal batteries nor are they equipped with individual audible alarms. Instead, they are daisy-chained along two-wire zones, with both low-voltage conductors isolated from ground, although usually in grounded

FIGURE 10-1 Fire alarm control panel. (*Wikipedia*)

FIGURE 10-2 Residential-type smoke detector. (*Wikipedia*)

metal raceway. The pair of wires performs four functions: it notifies the control panel if the wires become shorted to one another or to ground or become open, it notifies the control panel if one or more of the alarm heads detects products of combustion, and it notifies the control panel if a head becomes inoperative. The fourth function performed by the two conductors is that they supply power to the heads, as needed to bias the semiconductors, detect products of combustion, and energize the LED indicator (the LEDs remain on steady when they sense a fire).

The older fire alarm systems required personnel when the system went into alarm to walk the length of the zone and find the affected head. Newer central fire alarm systems have addressable heads, which report to the control panel and display in the alphanumeric readout the exact location of the event, which can be critical in a quickly propagating fire.

The heads are wired in parallel across the pair of zone wires. The heads are built so that normally they are open (high ohms resistance) and when detecting products of combustion they conduct (low ohms resistance). How then, you may ask, is the control panel able to detect an open zone without manually shunting out the final head? The answer is by means of an ingenious device known as an "end of line resistor." A low-watt resistor having a typical resistance of 47,000 ohms (varying for different manufacturers) is placed in parallel across the two-wire circuit after the final head, so that the control panel can perform a constant continuity check. This is an example of the system's supervisory function. In the U.S., the end-of-line resistor is located inside the junction box to which the base of the final head is bolted. Canada requires a separate enclosure. In Europe, an end-of-line capacitor is used.

The heads are known as *initiating devices*. There are other types as well, such as manual pull stations and water flow indicators so that the system will go into alarm, activating notification appliances, shown in Figure 10-3, if a sprinkler head opens. In this sense, each of the hundreds or thousands of sprinkler heads is an initiating device.

FIGURE 10-3   Fire alarm siren. (*Wikipedia*)

Another fundamental part of the fire alarm system consists of notification appliances, most prominently very loud horns located throughout the building, and bright strobe lights for the benefit of hearing-impaired individuals. Additionally, there are live or pre-recorded instructions and brightly illuminated message displays.

The defining element in a central fire alarm system is the control panel, often located in a business office or reception area. (There may also be remote panels, for example, in maintenance areas.) The control panel, usually in a large, permanently mounted steel cabinet with a hinged glass front cover, is the brain center of the otherwise highly decentralized fire alarm system. In the bottom section of the cabinet are two 12-volt batteries connected in series to supply 24 volts in case there is a power outage and the emergency system does not come on line immediately. There should not be an interruption of fire alarm coverage at any time.

The control panel performs many functions. In addition to activating the notification appliances when an initiating device detects products of combustion, the control panel contains a module to which two redundant telephone lines are terminated. If the system goes into alarm, an automated call is placed to a specified number, generally the fire department. These lines are automatically tested monthly, and if one of them is inoperative, the maintenance department is informed.

The fire alarm system at any given time is in one of three states—normal, trouble, and alarm. As just discussed, when the system goes into alarm, all notification appliances respond and building occupants take action as required. Since the fire alarm system reacts swiftly in the earliest stages of a fire, the expectation is that occupants will be safe and firefighters will take action so that lives will not be lost and property damage is minimized.

As previously stated, one category of notification device is the waterflow indicator, which detects water moving through the piping that supplies the sprinkler heads. This is an example of the ways in which building systems interact in the event of fire.

A point we have to be aware of when installing and maintaining elevator systems, as they relate to sprinklers, is where the heads are required, and where they are prohibited. We will begin, however, with an overview of the sprinkler system. Broadly speaking, there are two types, wet and dry. In a wet system, the pipes feeding each and every head are filled with water, fully pressurized, and ready to flood the area when the element in any sprinkler head melts out.

The elements are composed of various metal alloys with different melting points, determined by ceiling heights and type of fire anticipated. The dry system is more complex and costly to implement. Also, there is a delay before water is released. However, it has the advantage that water in the zone pipes is not subject to freezing, and therefore the dry system can be used in an unenclosed porch in cold regions.

Wet and dry sprinkler systems alike are supplied by large-capacity (often six-inch) pipes from the municipal system or local reservoir. This is necessary to furnish the high volume of water needed to extinguish a fire. Outside portions of the pipe are buried below worst-case frost level. In the wet system this pipe branches into two-inch segments, which supply the many heads in each zone. In the dry system, the water main also passes under the concrete foundation footing and elbows up to a large cast-iron or forged steel structure. Inside is a semi-flexible, spring-loaded flapper that is normally closed, preventing the outside water from entering the interior portion of the system. The high water pressure outside is balanced by the spring combined with pressurized air in the interior system, holding the flapper in the closed position.

Pressure gauges enable maintenance workers to monitor air pressure throughout the interior zones. They check each zone typically twice a day and record readings in a log so that long-term trends can be detected, and where necessary they add a few pounds of air by opening a small air compressor valve. If the air pressure goes below a user-determined level, the fire alarm system goes into the trouble state, and if water flow is sensed, it goes into the alarm state. For maintenance workers, an objective is to ensure that a trouble state will not degenerate into a fire alarm state, while seeing to it that a true alarm state is not suppressed.

Sprinkler heads are required in most jurisdictions to be installed at the top of the hoistway and in each machine room. Electrical disconnects are to be installed so as to shut off power in these areas prior to water flow to prevent erratic elevator car and door operation if electrical controls and wiring are flooded. The disconnects are actuated by heat detectors that operate at a lower temperature than the sprinkler heads.

In the event that a zone must be temporarily shut down for repairs, maintenance workers or security personnel are required to manually patrol the area watching for fire.

We have seen how the fire alarm system is structured to interact with the sprinkler system. Now we will discuss the highly structured protocols that govern the fire alarm system connection to the elevator system.

In the event of fire in a commercial building such as hotel, multi-story restaurant, or office building, large numbers of untrained individuals may assume that the way to safety is a passenger elevator. This is emphatically not the case. Stairways are enclosed within fire-rated walls and doors that close when the fire alarm system goes into the alarm state. Building occupants can best get to the ground floor via stairs.

Moreover, the elevator system enters the fire service mode, which is designed not to respond to the call button at each floor, but instead to proceed immediately to the ground or other designated floor and remain there with doors open until firefighters arrive and require the use of the elevator.

If firefighters occupy the car and need to exit at a specific floor, after the car stops with the door remaining closed, they will touch it. If it is hot, they will know not to open the door because there is fire on the other side of the door, but to take the car elsewhere.

Elevator fire service is divided into two parts: Phase I Emergency Recall and Phase II Emergency In-Car Operation. Phase I is initiated by smoke detectors in the elevator lobby, a landing, or the machine room. It is important that these smoke detectors are chosen and located so as to activate prior to nearby sprinkler heads. The smoke detectors are generally wired to activate the central fire alarm system and thereby notify the local fire department and/or other designated services.

Phase I Emergency Recall takes over the elevator system and causes the car to proceed to the lobby or designated alternate landing and to remain there with doors open. Arriving on the scene, firefighters have the option of initiating Phase II Emergency In-Car Operation. To do so, they again use their key to select firefighter operation. The key switch, with its familiar red light and fire helmet graphic, is located adjacent to the hoistway door. Once activated, the switch enables firefighters to open and close doors and move the car among floors, using buttons on the car control panel. Firefighters then use the car to access the fire location and to evacuate any remaining individuals who did not use the stairs.

If there is smoke in the hoistway or machine room, smoke detectors cause the fire helmet icon to flash, indicating that total elevator shutdown is imminent. Instructions posted adjacent to the in-car control panel state: "WHEN FIRE HELMET FLASHES, EXIT ELEVATOR." If elevator power is interrupted, it can only be reset manually in the machine room, otherwise the car may be stuck between floors. If they have been compliantly wired, communication and lights in the car remain functional since they are on dedicated circuits.

Emergency Power

Electric utilities contract to endeavor to provide continuous power, but they are not liable for damages incurred by customers due to loss of power. It is a recognized fact that there are outages, often weather-related or caused by unanticipated failure of transformers, switch gear, or other equipment. Because outages can be lengthy, lasting hours or even days, all but the smallest commercial facilities have backup power, usually one or more onsite generators. This is universally true of buildings having elevators. Small generators may be propane or natural gas powered, and larger ones invariably have diesel powered prime movers. In an outage, emergency lights with individual batteries light up immediately, and backup generators are required to be up to speed and online within 10 seconds. As part of a preventive maintenance program, elevator technicians should verify that emergency power is in place and functional at all times, and that routine maintenance is being performed. They should also make sure that replacement parts such as V-belts and fuel filters are in stock.

The National Electrical Code contains requirements for emergency systems in Article 700. It is not expected that every load supplied by the normal electrical system be connected to the emergency electrical system. Accordingly, the emergency system can be considerably smaller than the normal system. It must, however, be capable of supplying all essential loads, except for those that do not run simultaneously, such as heat and air-conditioning.

Like other NEC articles, Article 700 begins with definitions:

- **Branch-Circuit Emergency Lighting Transfer Switch.** A device connected on the load side of a branch-circuit over-current protective device that tranfers only emergency lighting loads from the normal supply to an emergency supply.

- **Emergency Systems**. Those systems legally required and classed as emergency by municipal, state, federal, or other codes or by any governmental agency having jurisdiction. These systems are intended to automatically supply illumination, power, or both to designated areas and equipment in the event of failure of the normal supply or in the event of accident to elements of a system intended to supply, distribute, and control power and illumination essential for safety to human life. Emergency systems are generally installed in places of assembly where artificial illumination is required for safe exiting and for panic control in buildings subject to occupancy by large numbers of persons, such as hotels, theaters, sports arenas, health care facilities, and similar institutions. Emergency systems may also provide power for such functions as ventilation where essential to maintain human life, fire detection and alarm systems, elevators, fire pumps, public safety communications systems, industrial processes where current interruption would produce serious life safety or health hazards, and similar functions.
- **Luminaires, Directly Controlled**. (Luminaire is the NEC term for light fixture.) An emergency luminaire that has a control input for an integral dimming or switching function that drives the luminaire to full illumination upon loss of normal power.
- **Relay, Automatic Load Control**. A device used to set normally dimmed or normally off switched emergency lighting equipment to full power illumination levels in the event of a loss of the normal supply by bypassing the dimming/switching controls, and to return the emergency lighting equipment to normal status when the device senses that the normal supply has been restored.

Tests and Maintenance

The authority having jurisdiction (this is the NEC term for electrical inspector) is to conduct or witness a test of the complete system upon installation and periodically afterward. Emergency systems are to be tested periodically on a schedule acceptable to the authority having jurisdiction to ensure the systems are maintained in proper operating condition. Emergency system equipment is to be maintained in accordance with manufacturers' instructions and industry standards. A written record is to be kept of such tests and maintenance. Means for testing all emergency lighting and power systems during maximum anticipated load are to be provided. (During a utility outage, local maintenance workers can use the opportunity to perform these tests.)

If the emergency system relies on a single alternate source of power, which will be disabled for maintenance or repair, the emergency system is to include permanent switching means to connect a portable or temporary alternate source of power, which is to be available for the duration of the maintenance or repair. The permanent switching means to connect a portable or temporary alternate source of power is to comply with the following:

- Connection to the portable or temporary alternate source of power is not to require modification of the permanent system wiring.
- Transfer of power between the normal power source and the emergency power source is to be in accordance with Section 700.12, General Requirements for Sources of Power (see below).
- The connection point for the portable or temporary alternate source is to be marked with the phase rotation and system bonding requirements.
- Mechanical or electrical interlocking is to prevent inadvertent interconnection of power sources.
- The switching means are to include a contact point that annunciates at a location remote from the generator or at another facility monitoring system to indicate that the permanent emergency source is disconnected from the emergency system.

It is permissible to use manual switching to switch from the permanent source of power to the portable or temporary alternate source of power and to use the switching means for connection to a load bank.

Capacity and Rating. An emergency system is to have adequate capacity and rating for all loads to be operated simultaneously. The emergency system equipment is to be suitable for the maximum fault current at its terminals.

Selective Load Pickup, Load Shedding, and Peak Load Shaving. The alternate power source is permitted to supply emergency, legally required standby and optional standby system loads where the source has adequate capacity or where automatic selective load pickup and load shedding are provided as needed to ensure adequate power to (1) emergency circuits, (2) the legally required standby circuits, and (3) the optional standby circuits, in that order of priority. The alternate power source is permitted to be used for peak load shaving, provided these conditions are met.

Peak load shaving operation is permitted for satisfying the test requirement for periodic tests.

Transfer Equipment, General. Transfer equipment, including automatic transfer switches, is to be automatic, identified for emergency use, and approved by the authority having jurisdiction. Transfer equipment is to be designed and installed to prevent the inadvertent connection of normal and emergency sources of supply in any operation of the transfer equipment. Transfer equipment and electric power production systems installed to permit operation in parallel with the normal source are to meet the requirements of Article 705, Interconnected Electric Power Production Sources.

Bypass Isolation Switches. Means are permitted to bypass and isolate the transfer equipment. Where bypass isolation switches are used, inadvertent parallel operation is to be avoided.

Automatic Transfer Switches. Automatic transfer switches are to be electrically operated and mechanically held. Automatic transfer switches are to be listed for emergency system use.

Use. Transfer equipment is to supply only emergency loads.

Documentation. The short-circuit current rating of the transfer equipment, based on the specific over-current protective device type and settings protecting the transfer equipment, is to be field marked on the exterior of the transfer equipment.

Signals. Audible and visual signal devices are to be provided, where practicable, for the following purposes:

- Malfunction—to indicate malfunction of the emergency source.
- Carrying Load—to indicate that the battery is carrying load.
- Not Functioning—to indicate that the battery charger is not functioning.
- Ground Fault—to indicate a ground fault in solidly-grounded wye emergency systems of more than 150 volts to ground and circuit protective devices rated 1000 amperes or more. The sensor for the ground-fault signal devices is to be located at, or ahead of, the main system disconnecting means for the emergency source and the maximum setting of the signal devices is to be for a ground-fault current of 1200 amperes. Instructions on the course of action to be taken in the event of indicated ground fault are to be located at or near the sensor location. For systems with multiple emergency sources connected to a paralleling bus, the ground fault sensor is permitted to be at an alternate location.

Signs. A sign is to be placed at the service entrance equipment, indicating type and location of each on-site emergency power source.

Where removal of a grounding or bonding connection in normal power source equipment interrupts the grounding electrode conductor connection to the alternate power source(s) grounded conductor, a warning sign is to be installed at the normal power equipment stating:

WARNING!

SHOCK HAZARD EXISTS IF GROUNDING

ELECTRODE CONDUCTOR OR BONDING JUMPER

CONNECTION OF THIS EQUIPMENT IS REMOVED

WHILE ALTERNATE SOURCE(S) ARE ENERGIZED

Surge Protection. A listed surge protective device is to be installed in or on all emergency systems switchboards and panelboards.

Wiring, Emergency System Identification. Emergency circuits are to be permanently marked so they will be readily identified as a component of an emergency circuit or system by the following methods:

- All boxes and enclosures (including transfer switches, generators, and power panels) for emergency circuits are to be permanently marked as a component of an emergency system.

- Where boxes or enclosures are not encountered, exposed cable or raceway systems are to be permanently marked to be identified as a component of an emergency circuit or system, at intervals not to exceed 25 feet.
- Receptacles supplied from the emergency system are to have distinctive color or marking on the receptacle cover plates or receptacles.

Wiring. Wiring of two or more emergency circuits supplied from the same source are permitted in the same raceway, cable, box, or cabinet. Wiring from an emergency source or emergency source distribution over-current to emergency loads is to be kept entirely independent of all other wiring and equipment, unless otherwise permitted as follows:

- Wiring from normal power source located in transfer equipment enclosures.
- Wiring supplied from two sources in exit or emergency luminaires.
- Wiring from two sources in a listed load control relay suppling exit or emergency luminaires, or in a common junction box, attached to exit or emergency luminaires.
- Wiring within a common junction box attached to unit equipment, containing only the branch circuit supplying the unit equipment and the emergency circuit supplied by the unit equipment.
- Wiring from an emergency source to supply emergency and other loads in accordance with 1-4 as follows:

1. Separate vertical switchgear sections or separate vertical switchboard sections, with or without a common bus, or individual disconnects mounted in separate enclosures are to be used to separate emergency loads from all other loads.

2. The common bus of separate sections of the switchgear, separate sections of the switchboard or the enclosure are to be either of the following:
    - Supplied by a single or multiple feeders without over-current at the source
    - Supplied by a single or multiple feeders with over-current protection, provided that the over-current protection that is common to an emergency system and any non-emergency systems(s) is selectively coordinated with the next downstream over-current protective device in the non-emergency systems.
    - Emergency circuits are not to originate from the same vertical switchgear section, vertical switchboard section, panelboard enclosure, or individual disconnect enclosure as other circuits.
    - It is permissible to use single or multiple feeders to supply distribution equipment between an emergency source and the point where the emergency loads are separated from all other loads.

3. Wiring Design and Location. Emergency wiring circuits are to be designed and located so as to minimize the hazards that might cause failure due to flooding, fire, icing, vandalism, and other adverse conditions.

4. Fire Protection. Emergency systems are to meet the additional requirements in the following occupancies:

- Assembly occupancies for not less than 1000 persons
- Buildings above 75 feet in height
- Health care occupancies with more than 300 occupants

Feeder-Circuit Wiring. Feeder-circuit wiring is to meet one of the following conditions:

- The cable or raceway is installed in spaces or areas that are fully protected by an approved automatic fire suppression system.
- The cable or raceway is protected by a listed electrical circuit protective system with a minimum two-hour fire rating.
- The cable or raceway is a listed fire-resistive cable system.
- The cable or raceway is protected by a listed fire-rated assembly that has a minimum fire rating of two hours and contains only emergency circuits.
- The cable or raceway is enclosed in a minimum of two inches of concrete.

Feeder-Circuit Equipment. Feeder-circuit equipment for feeder circuits (including transfer switches, transformers, and panelboards) is to be located either in spaces fully protected by approved automatic fire-suppression systems (including sprinklers or carbon dioxide systems) or in spaces with a two-hour fire-resistive rating.

Generator Control Wiring. Control conductors installed between the transfer equipment and the emergency generator are to be kept entirely independent of all other wiring and are to meet the conditions of 700.10(D)(1) for feeder circuit wiring. The integrity of the generator control wiring is to be continuously monitored. Loss of integrity of the remote start circuit(s) is to initiate visual and audible annunciation of generator malfunction at the generator local and remote annunciator(s), and start the generator(s).

Sources of Power. General Requirements: current supply is to be such that in the event of failure of the normal supply to or within the building, emergency lighting, emergency power, or both are to be available within the time required by the application, but not to exceed 10 seconds. The supply system for emergency purposes, in addition to the normal services to the building is to be one of the following:

- **Storage Battery.** Storage batteries are to be of suitable rating and capacity to supply and maintain the total load for a minimum period of 1½ hours, without the voltage applied to the load falling below 87½ percent of normal. Automotive-type batteries are not to be used. An automatic battery charging means is to be provided.

- **Generator Set, Prime-Mover Driven.** For a generator set driven by a prime mover acceptable to the authority having jurisdiction, means are to be provided for automatically starting the prime mover on failure of the normal service and for automatic transfer and operation of all electric circuits. A time delay, permitting a 15-minute setting, is to be provided to avoid retransfer in case of short-time re-establishment of the normal source.

- **Internal Combustion Engines as Prime Mover.** Where internal combustion engines are used as the prime mover, an on-site fuel supply is to be provided with on-premises fuel supply sufficient for not less than two hours full demand operation of the fuel transfer pumps to deliver fuel for a generator set day tank. This pump is to be connected to the emergency power system.

Dual Supplies. Prime movers are not to be solely dependent on a public utility gas system for their fuel supply or municipal water supply for their cooling systems. Means are to be provided for automatically transferring from one fuel supply to another where dual fuel supplies are used.

Battery Power and Damper. Where a storage battery is used for control of signal power or as the means for starting the prime mover, it is to be suitable for the purpose and is to be equipped with an automatic charging means independent of the generator set. Where the battery charger is required for the operation of the generator set, it is to be connected to the emergency system. Where power is required for the operation of dampers used to ventilate the generator set, the dampers are to be connected to the emergency system.

Auxiliary Power Supply. Generators that require more than 10 seconds to develop power are permitted if an auxiliary power supply energizes the emergency system until the generator can pick up the load.

Outdoor Generator Set. Where an outdoor housed generator set is equipped with a readily accessible disconnecting means and the disconnecting means is located within sight of the building supplied, an additional disconnecting means is not required where ungrounded conductors serve or pass through the building.

Fuel Cell System. Fuel cell systems used as a source of power for emergency systems are to be of suitable rating and capacity to supply and maintain the total load for not less than two hours of full-demand operation. Where a single fuel cell system serves as the normal supply for the building, it is not to serve as the sole source of power for the emergency standby system.

Emergency System Circuits for Lighting and Power/Loads on Emergency Branch Circuits. No appliances and no lamps other than those specified as required for emergency use are to be supplied by emergency circuits.

Emergency Illumination is to include means of egress lighting, illuminated exit signs, and all other luminaires specified and necessary to provide required illumination. Emergency lighting systems are to be designed and installed so that the failure of any individual lighting element, such as the burning out of a lamp, cannot leave in total darkness any space that requires emergency illumination. Where high intensity discharge lighting such as high- and low-pressure sodium, mercury vapor and metal halide is used as the sole source of normal illumination, the emergency lighting system is required to operate until normal illumination has been restored.

Where an emergency system is installed, emergency illumination is to be provided in the area of the disconnecting means, where the disconnecting means are installed indoors.

Branch Circuits for Emergency Lighting. Branch circuits that supply emergency lighting are to be installed to provide service where the normal supply for lighting is interrupted. Such installations are to provide for either of the following:

- An emergency lighting supply, independent of the normal lighting supply, with provisions for automatically transferring the emergency lights upon the event of failure of the normal lighting branch circuit.
- Two or more branch circuits supplied from separate and complete systems with independent power sources. One of the two power sources and systems is to be part of the emergency system and the other is permitted to be part of the normal power source and system. Each system is to provide sufficient power for emergency lighting purposes.

Unless both systems are used for regular lighting purposes and are both kept lighted, means is to be provided for automatically energizing either system upon failure of the other. Either or both systems are permitted to be part of the general lighting of the protected occupancy.

Circuits for Emergency Power. For branch circuits that supply equipment classed as emergency, there shall be an emergency supply source to which the load will be transferred automatically upon failure of the normal supply.

Control: Emergency Lighting Circuits Switch Requirements. The switch or switches installed in emergency lighting circuits are to be arranged so that only authorized persons have control of the emergency lighting.

Switch Location. All manual switches for controlling emergency circuits are to be in locations convenient for authorized persons responsible for their activation.

Normal Electrical System

Most electrical utilities maintain service to their customers on a near continuous basis, with local emergency systems taking over a very small percentage of the time. When they do come online, it is often in response to a fault in the customer's equipment as opposed to a utility outage.

The normal electrical system in a typical commercial facility is online and fully functional better than 95 percent of the time. It's worth taking a look at how this system supplies the elevators in regard to the electrical connections. Except for exceptional circumstances, a building has only one electrical service. This is defined as the conductors and equipment for delivering electric energy from the serving utility to the wiring system of the premises served. Generally, it is considered to extend from the utility point of connection to the customer's first (farthest upstream) over-current device or main breaker. This is usually installed in the entrance panel, although alternatively it can be in a separate enclosure. Either of these cabinets can be indoors or, if weatherproof, outdoors.

The point of connection is determined by utility policy. It can be a few inches upstream from the weatherhead, allowing for a drip loop, or it can be at the meter input lugs. For an underground service, the point of connection is usually at the meter. Actually, the meter location, despite what many believe, does not define anything. It can be on the power pole, outdoors on a pedestal, or mounted on the building or indoors in a utility room.

The first, and very critical, stage in doing the electrical design for a new occupancy is sizing out the service. If it is oversized, that is a waste of resources except where future expansion is envisioned. If it is undersized, there is the prospect of the main breaker tripping out or even of an electrical fire.

To size out an electrical service, the type of occupancy and its size in square feet are considered in determining the lighting load. Then, any connected electrical loads are added. The smaller of two non-simultaneous loads, such as heating and air-conditioning, may be omitted. Continuous loads, expected to run more than three hours, must be increased to 125 percent. This same factor applies in most cases to motors. Connected loads are permitted to be derated by differing factors according to the nature of the loads. All loads, including lighting, are figured in watts. After derating factors are applied, the total load is divided by the number of volts applied to the loads in accordance with Watt's law and the result in amperes is used to size the service, electrical conductors, over-current devices, and other devices such as switches and receptacles.

All this is laid out by the NEC in great detail, together with exceptions, tables, diagrams, and notes.

Ordinary residences and small to medium-sized commercial facilities are usually designed by the electrician, while larger and more complex projects are designed by professional engineers experienced in the specialized field. In large projects, efficient data cabling and lighting are highly specialized, sometimes massive undertakings requiring the services of experts.

Where present, an elevator system is often the greatest single electrical load, due to the large motor(s).

NEC Article 430, Motors, Motor Circuits and Controllers, is one of the longest articles, with numerous subsections. Motors are exceptionally diverse, compared to most electrical equipment. In large motors, mechanical and electrical forces are powerful and potentially hazardous. In elevator systems, the motors must perform as expected, or users may be inconvenienced, injured, or worse.

Because of the nature of the equipment, elevator technicians must look closely at the electrical motors. Errors in design and initial installation are not always apparent at startup, and even the best testing and evaluating may fail to reveal problems that may surface later on.

If the initial installation was undertaken by the manufacturer or by a reputable independent installation firm, the end product should be efficient, reliable, and free of hazards. As always, however, elevator maintenance technicians, whether regular employees at the facility or brought in on a case by case basis, should continuously examine all aspects of the installation to see if it is compliant.

If the motor is an original component, supplied by the manufacturer of the elevator system, it is generally safe to assume that it has been sized out correctly and is suitable for the application. At the time of installation and prior to being placed in service, it is also generally safe to assume that the electrical supply wiring and over-current protective devices have been sized out correctly. In the interest of getting this right, the facility elevator technicians, working with the electricians, should review NEC Article 430 to determine if the elevator installation in regard to the motor(s) is compliant.

To begin, review Section 430.6, Ampacity and Motor Rating Determination, which states that the size of conductors supplying equipment covered by Article 430 is to be selected from the allowable ampacity tables in accordance with Section 310.15(B) or to be calculated in accordance with Section 310.15(C). The required ampacity and motor ratings are to be determined as follows:

Other than for motors built for low speeds (less than 1200 RPM) or high torques and for multispeed motors, the values given in Table 430.247, Table 430.248, Table 430.249 and Table 430.250 are to be used to determine the ampacity of conductors or ampere rating of switches and branch-circuit short-circuit and ground-fault protection, instead of the actual current rating marked on the motor nameplate.

This is indeed a curveball, and the wording and implications must be understood before going further. Most appliances, portable and stationary power tools, and all but the smallest motors have metal nameplates with permanent lettering inscribed. To size out the electrical circuits including over-current protection for these loads, you take the current rating off the nameplate and, consulting NEC tables in Section 310.15(B) you choose the correct conductors to supply the equipment.

In contrast to this procedure, to size out the conductors for most motors, you first determine the motor horsepower, also given on the nameplate. Then, consulting the four tables cited above (430.247-250) and ignoring the current values on the motor nameplate, you take the correct full load current. These four tables are found at the end of Article 430, and must be consulted to size out the circuits supplying most motors.

Wiring Methods for Elevator Shafts, Pits, and Machine Rooms

Type NM (Non-metallic sheathed, trade name Romex) is used almost universally in residential and light commercial occupancies. It is inexpensive, quick and easy to install, and safe in those settings. But it should not be seen in buildings over three stories, commercial garages, industrial facilities, and similarly demanding locations.

Romex is absolutely non-compliant in elevator systems. Instead, you will usually see Type EMT (electrical metallic tubing) and Type MC (metal-clad) cable. In electrical alterations or expansion, for example, extending a branch circuit to add an extra light fixture in a machine room, these two wiring methods are the best choice.

EMT is not, properly speaking, a type of conduit. It is tubing, although electricians, when referring to it, speak of "putting the wire in conduit." It is plenty rugged for elevator applications, unless they happen to be in a classified (hazardous) location where there are or may be flammable gases, vapors from hazardous liquids, or combustible dust particles suspended in the air. In these extremely sensitive locations, Type RMC (rigid metal conduit) is generally required. RMC is heavy steel pipe. In dimensions and threads, it is equivalent to galvanized steel water pipe, but water pipe must never be used in place of conduit in electrical installations. RMC is far more costly than EMT. Power cutters and threaders are required, and the work is strenuous and time consuming.

Small diameter (½", ¾" and 1") EMT cuts readily with a hacksaw and the fittings slide onto the pipes and fasten with simple setscrews. Angle bends are made using an inexpensive hand bender. The bends have the correct radius for each size EMT so conductors can be easily pushed through. If sharp angles are needed, as when two walls join to make an exterior corner, pull fittings with removable covers are available so that the wire can be pulled to that point, out through the opening in the fitting, then re-inserted to continue the installation. As in all raceway systems, NEC requires that each pipe run is to be completed with both ends terminated at their respective enclosures prior to installing the conductors. (You can't pull the wires as you assemble the raceway.)

For simple short runs, the wire can be pushed through. For longer runs with more bends, a pull rope is necessary. Unlike the conductors, the pull rope can be inserted into each piece of EMT and all fittings as they are assembled. To facilitate, since the EMT lengths are 10 foot, tape the end of the pull rope to a thin rod slightly longer and poke it through.

If you decide a pull rope is needed after the EMT run is done, use a tool known as a piston, or mouse, with fishline attached. Then use a shop vac at the far end to pull it through. The fishline is then used to draw a full-sized pull rope through. If you don't have the right size piston, carve one out of Styrofoam.

For most applications where redundant grounding is not needed, EMT qualifies as the equipment grounding conductor. However, most good electricians "pull a green for everything" and terminate it at both ends.

Full installation requirements, uses permitted and uses not permitted, are found in NEC Article 358. Some points to remember:

- EMT may be exposed or concealed in dry, damp or wet locations. For outdoor installations, special compression fittings are used rather than the setscrew type.

- EMT is permitted and available in trade sizes ½-inch through 4-inch. In the larger sizes, a power bender is needed.
- There are not to be more than the equivalent of four quarter bends (360 degrees total) between pull points; for example, conduit bodies and boxes.
- All cut ends of EMT are to be reamed to remove rough edges.

Most of the time, THHN conductors are used in the appropriate ampacity, 12 AWG for small branch circuits. For some applications, Romex can be pulled through metal raceway, but it is costly, quickly overfills the conduit, and although Code compliant, generally is considered bad form. In all cases, maximum conduit fill rules must be observed. While not usually required for data and telephone wiring, EMT makes a good installation between floors in a building, and if future alterations are desired, new wires can be easily pulled.

As you can see from the above, a short, direct EMT run is quite simple. For a complex run of many pipes, especially where some runs make a turn and others go straight, and where looks are important because the runs will be exposed to view as on a ceiling, the work becomes exacting and there are challenges. The way to master this interesting subtrade is to work with an experienced electrician in quite a few jobs.

In a large, complex installation, some sections cannot easily be done in EMT. Where the run has many odd bends at close intervals, it has to go through cavities behind finished walls, or it must go through drilled holes in framing members, EMT is not suitable. Fortunately, there is an alternative. It is Type MC (metal clad) cable. This is a flexible metal jacketed cable that resembles the old BX, which is no longer used. MC comes in rolls of various lengths. The electrical supplier will cut just the length needed. MC also resembles Type AC (armored cable). It looks the same from the outside, but rather than a green equipment-grounding conductor, it contains a metal strip in contact with the armor so that the metal shield qualifies as an equipment ground. For elevator work, stay away from Type AC.

Type MC has pre-installed conductors, so unlike flexible metal conduit, which also looks the same from the outside, you don't have to pull wires. In placing an order, ask for something like Type MC 12-2. Besides the two current-carrying wires (black and white), there is a green equipment-grounding conductor.

Begin the installation by cutting a piece of MC to the correct length, adding extra for the required free conductors at both ends. The exact amount varies with the enclosure situation. This cut can be made with a hacksaw or large shears. It doesn't matter if you pinch the armor because these pieces will be removed and discarded, leaving the free conductors.

Where the short pieces of armor are cut at either end so that they can be slid off, you have to use a special tool known as an MC cutter (trade name Roto-Splitter). This sells for a little over $30 at Amazon.com, or it can be purchased from any electrical supplier. The tool clamps onto the cable, and when you turn the crank, a small circular blade cuts through the armor with no danger of nicking the enclosed insulated conductors. Then, slide off the scraps of armor and remove the plastic strips.

The MC fastens securely by means of an MC connector, which resembles a Romex connector but clamps onto the round armor. Be sure to insert the insulating bushing, which prevents the conductors from chafing. A quick and easy alternative is the snap-on connector, which clicks into the knockout with no bushing or locknut required.

Installation requirements with Uses Permitted and Uses Not Permitted are found in NEC Article 330. Here are some key points:

- MC is permitted for services, feeders, and branch circuits.
- MC is permitted indoors or outdoors.
- MC is permitted exposed or concealed.
- MC is to be secured at intervals not exceeding six feet.
- Cables containing four or fewer conductors sized no larger than 10 AWG are to be secured within 12 inches of every box or other cable termination.

Uses Permitted and Uses Not Permitted are similar to those for Type EMT, so for the most part they can be used interchangeably, and both are suitable for most elevator work. The good news is that it is easy to transition from one to the other. Terminate EMT and MC at a 4 × 4 square box or other suitable metal enclosure, and connect the conductors using wire nuts. Don't forget to ground the box.

STUDY QUESTIONS

1. A centrally controlled and automatically monitored fire alarm system:
   A. is attended by a human at all times
   B. sends and receives signals from the notification appliances
   C. activates sprinkler heads throughout the building
   D. dispatches persons to check every room in the building

2. When the sprinkler system discharges water:
   A. the fire alarm system goes into the alarm state
   B. all detector heads are activated
   C. all sprinkler heads are opened simultaneously
   D. the fire station is notified that it may receive a call later

3. When activated:
   A. the LEDs remain on steady when they sense a fire
   B. the sprinkler system closes all fire doors
   C. the fire alarm system disconnects the telephone system
   D. the sprinkler system disconnects the telephone system

4. In a centrally monitored and controlled fire alarm system:
   A. the individual heads have batteries
   B. the individual heads have loud annunciators
   C. the LEDs remain on steady when they sense a fire
   D. the individual heads have LEDs that blink when the heads sense products of combustion

5. Heads are wired _____ to the central control panel.
   A. in series
   B. in parallel
   C. by means of three-wire circuits
   D. at 120 volts

6. The fire alarm heads are:
   A. initiating devices
   B. notification appliances
   C. filled with a heat-sensitive liquid
   D. difficult to replace

7. Pull stations:
   A. cannot be re-used
   B. are activated by humans
   C. are wired with the annunciators
   D. contain internal batteries

8. Elevator Fire Service consists of:
   A. Phase I Emergency Recall
   B. Phase II Emergency In-Car Operation
   C. both of the above
   D. neither of the above

9. Emergency backup generators are required to be up to speed and online within:
   A. ten seconds
   B. twenty seconds
   C. thirty seconds
   D. sixty seconds

10. Emergency system transfer switches:
   A. are fully automatic
   B. are operated by humans
   C. keep utility lines energized during an outage
   D. are not covered in the National Electrical Code

For answers, go to Appendix A.

STUDY QUESTIONS ANSWERS

Chapter 1

1. B
2. B
3. C
4. D
5. A
6. D
7. B
8. A
9. A
10. D

Chapter 2

1. C
2. D
3. B
4. A
5. B
6. C
7. A
8. A
9. D
10. C

Chapter 3

1. B
2. D
3. D
4. A
5. C
6. D
7. A
8. B
9. D
10. D

Chapter 4

1. A
2. B
3. D
4. A
5. B
6. A
7. C
8. D
9. A
10. B

Chapter 5

1. C
2. D
3. B
4. D
5. A
6. D
7. A
8. D
9. B
10. C

Chapter 6

1. A
2. B
3. C
4. B
5. D
6. D
7. D
8. D
9. B
10. D

Chapter 7

1. B
2. C
3. A
4. C
5. D
6. D
7. A
8. B
9. C
10. D

Chapter 8

1. B
2. C
3. A
4. A
5. D
6. B
7. D
8. D
9. D
10. D

Chapter 9

1. D
2. A
3. B
4. B
5. D
6. B
7. B
8. C
9. A
10. C

Chapter 10

1. B
2. A
3. A
4. C
5. B
6. A
7. B
8. C
9. A
10. A

ELECTRICAL LAWS AND EQUATIONS

Ohm's Law Wheel, shown in Figure B-1, is a convenient circular tabulation that displays Ohm's Law and the Electrical Power Law. You can easily access solutions for volts, amps, ohms and power unknowns, rather than performing multiple transpositions.

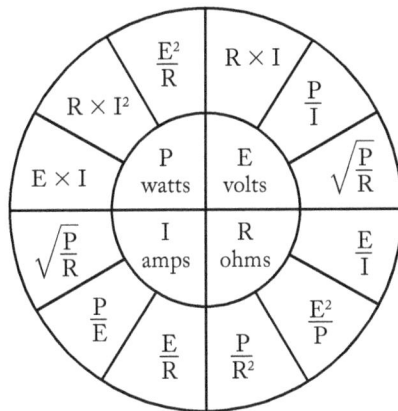

FIGURE B-1   Ohm's Law Wheel.

These electrical parameters are not empirical observations. They are fixed by definition. One volt, for example, is defined as the electrical potential that will cause one amp of current to flow through a one-ohm load. These relations exist in the real world in accordance with the following definitions:

One amp is the amount of current that flows through a conductor when 6.24×10^{18} electrons pass a given point on the conductor each second. This is furthermore the definition of a charge of one coulomb.

It's clear that R refers to resistance in ohms, but what about E and I? E signifies electromagnetic force, the technical term for voltage. I signifies intensity, an old word for current. Voltage is applied to a load and current flows through it. It makes no sense to say that 10 volts flows through a conductor or load, or that 10 amps is applied to it.

The most basic equation, $E = I \times R$, is applicable when voltage is applied to and current flow through a purely resistive load. Not all loads are purely resistive. Some loads are capacitive and some are inductive. Many (theoretically all) are partly resistive and partly capacitive or inductive. Capacitance and inductance are properties of a component, circuit or transmission line and they do not change unless some physical change takes place. In other words, a component will have a specific capacitance or inductance whether it is in packaging on a shelf or in an energized circuit.

Based on its capacitance or inductance, these components have capacitive and inductive reactance, measured in ohms. However, they do not have a fixed relationship to the amount of current as does resistance in a purely resistive load. Instead, capacitive and inductive are frequency dependent. At higher frequencies given the same current, capacitive reactance is less and inductive reactance is greater. (To calculate the effect of these components, of course, you also have to look at whether they are connected is series or parallel to the load.)

The equation that governs capacitive reactance is:

$$X_C = 1/2\pi fC$$

where X_C = capacitive reactance
f = frequency
C = capacitance

Since frequency is in the denominator, capacitive reactance varies with it. It goes down as frequency goes up.

The equation for inductive reactance is:

$$X_L = 2\pi fL$$

where X_L = inductive reactance
f = frequency
L = inductance

Since f is no longer in the denominator, a rise in frequency causes the inductive reactance to go up.

INDEX

Page numbers followed by *f* and *t* refer to figures and tables, respectively.

Disconnecting means and control, 169–172,
179
Displacement switch, xiii
Dolivo-Dobrovolsky, Mikhail, 135
Door (gate) electric contacts, xiii
car, xii, xvii, 29, 89, 99, 186
faulty circuits in, 92–93, 189
hoistway, 29, 89, 99, 186
Double-deck elevators, 34
Down Peak Mode operating mode, 195
Driving machine, xiv, 85–87
connection to, 26, 95
location of, 173
operating, with open or unlocked doors,
91–92, 188
Driving-machine brake, xviii, 92, 188
Dumbwaiters, 35
Dynamic balancing, 69
Dynamos, 8

Edison, Thomas, 6–8, 10
Edwards, Bradley, 17
E/E/PE designation, xiii
Electric contacts:
car door/gate, xii, xvii, 29, 89, 99, 186
car-side emergency exit, 89, 186
faulty circuits in, 92–93, 189
for hinged car platform sills, xvii, 89,
186
hoistway door, 29, 89, 99, 186
Electric elevators:
ASME A17 on, xv–xviii, 85–95
grounding for, 173
history of, 10–11
Electric motors:
DC, 6–8, 13, 41, 64
early, 6–7, 7f
electrical load from, 212–213
maintaining (see Motor maintenance)
at Otis Brothers and Company, 10–11
overload protection for, 172
repairing (see Motor repair)
three-phase, 41, 52–54, 73, 137, 149–
151, 150f
(See also AC motors)
Electric shock, 50
Electrical diagnostic instruments, 109–132
for advanced elevator repair, 64
multimeters, 111–113
non-contact meters, 113–114

oscilloscopes, 116–129
PC-based spectrum analyzers, 130–131
power quality measurements with,
114–116
test lights, 109–111
thermal imagers, 131–132
Electrical distribution system, 6, 7
Electrical fuse, 102
Electrical issues, with motors, 49–51
Electrical metallic tubing (EMT), 164f,
214–216
Electrical Power Law, 221
Electrical protective devices, 29, 87–90,
98–99, 184–187
Electrical service, sizing, 212
Electrical system:
emergency power, 204–211
normal operation, 211–216
testing and maintenance for emergency
power system, 205–211
Electrical/electronic/programmable elec-
tronic (E/E/PE) designation, xiii
Electrician's license, x–xi
Electromagnetic force, 222
Elevator brake, 87, 175, 191
Elevator diagnostics field, 63
Elevator Group Control System
ΣAI-2200C, 37
Elevator manufacturers, 35–38
Elevator mechanic's license, x, xi
Elevator nonstop switch, xiv
Elevator rope, 158f
of early elevators, 1, 1f, 3, 4
in elevator system, 175
in high-rise buildings, 15–16
for traction elevators, 21, 23–24, 23f, 75
Elevator shaft, wiring methods, 213–216
Elevator systems, 85–106, 157–175
ASME A17 codes on, 85–101
components of, 174–175
error codes in service manuals, 102–104
maintenance for, 157–158
motion controller faults, 104–105
NEC requirements, 159–174
safety features in, 101–102
sensors in, 105–106
Elevators, Dumbwaiters, Escalators,
Moving Walks, Lifts, and Chairlifts
(Article 620, NEC), 159–174
conductors, 161–163, 167–168

ABOUT THE AUTHOR

After receiving a BA in English literature and composition from Hobart College, Geneva, New York, in 1965, David Herres did graduate work in philosophy at the New School for Social Research Graduate Faculty in New York City.

From 1970 to 1999, he owned and operated a construction company in Colebrook, New Hampshire, building homes and doing concrete and excavation work for other builders. He acquired a New Hampshire electrician's license in 1975 and did electrical installations from 2000 to the present.

He began writing articles on electrical work for *Electrical Construction and Maintenance* and other trade journals in 2006 and wrote four books on electrical work and electronics published by McGraw-Hill. He has written articles and CEUs for Jade Learning on the National Electrical Code and is contributing editor for *Design World Test and Measurement* and *Elevator World*.

He has written for many other publications including *Nut and Volts* and *Engineering News Record*. He has written a book on oscilloscopes, now in production at Springer Publishers and is currently writing a book on electronic instrumentation.

www.ingramcontent.com/pod-product-compliance
Lightning Source LLC
Chambersburg PA
CBHW061936190326
41458CB00009B/2755